# Student Solutions Manual

**Bob Martin**
*Tarrant County College*

Jamie Blair John Tobey Jeffrey Slater

# Prealgebra

## Third Edition

Upper Saddle River, NJ 07458

Editor-in-Chief: Chris Hoag
Executive Editor: Paul Murphy
Supplement Editor: Christina Simoneau
Executive Managing Editor: Kathleen Schiaparelli
Assistant Managing Editor: Becca Richter
Production Editor: Donna Crilly
Supplement Cover Manager: Paul Gourhan
Supplement Cover Designer: Joanne Alexandris
Manufacturing Buyer: Ilene Kahn

Pearson Prentice Hall
Pearson Education, Inc.
Upper Saddle River, NJ 07458



Printed in the United States of America

10 9 8 7 6 5 4 3 2 1

ISBN 0-13-149122-9 Standalone
0-13-186124-7 Student Study Pack Component

Pearson Education Ltd., *London*
Pearson Education Australia Pty. Ltd., *Sydney*
Pearson Education Singapore, Pte. Ltd.
Pearson Education North Asia Ltd., *Hong Kong*
Pearson Education Canada, Inc., *Toronto*
Pearson Educación de Mexico, S.A. de C.V.
Pearson Education—Japan, *Tokyo*
Pearson Education Malaysia, Pte. Ltd.

# Contents

# Chapter 1

## 1.1 Exercises

**1.** (a) 8002: Eight thousand two
(b) 802: Eight hundred two
(c) 82: Eighty-two
(d) In the number "eight hundred twenty" the place value of the digit "0" is one

**3.** (a) In the number 9865 the digit 8 is in the hundreds place.
(b) In the number 9865 the digit 5 is in the ones place.

**5.** (a) In the number 754,310 the digit 4 is in the thousands place.
(b) In the number 754,310 the digit 7 is in the hundred thousands place.

**7.** (a) In the number 1,284,073 the digit 1 is in the millions place.
(b) In the number 1,284,073 the digit 0 is in the hundreds place.

**9.** $4967 = 4000 + 900 + 60 + 7$

**11.** $2493 = 2000 + 400 + 90 + 3$

**13.** $867,301 = 800,000 + 60,000 + 7000 + 300 + 1$

**15.** 5 hundred-dollar bills, 6 ten-dollar bills, and 2 one-dollar bills.

**17.** (a) 4 ten-dollar bills and 6 one-dollar bills
(b) 4 ten-dollar bills, 1 five-dollar bill, and 1 one-dollar bill. Answers may vary.

**19.** 3125 Three thousand one hundred twenty-five

**21.** 42,125 Forty-two thousand, one hundred twenty-five

**23.** \$672.00
Six hundred seventy-two and 00/100

**25.** $1 < 7$

**27.** $4 < 8$

**29.** $11 > 10$

**31.** $9 > 0$

**33.** $2131 > 1909$

**35.** $52,647 < 616,000$

**37.** $5 > 2$

**39.** $2 < 5$

**41.** Expedition XLT > Supercab XLT

**43.** $45 = 50$ to nearest ten

**45.** $661 = 660$ to nearest ten

**47.** $16,462 = 16,500$ to nearest hundred

**49.** $823,042 = 823,000$ to nearest hundred

**51.** $38,431 = 38,000$ to nearest thousand

**53.** $12,577 = 13,000$ to nearest thousand

**55.** $5,254,423 = 5,300,000$ to nearest hundred thousand

**57.** $9,007,601 = 9,000,000$ to nearest hundred thousand

**59.** $865,000 = 870,000$ mi to nearest ten thousand

**61.** $16,962 = 17,000$ to nearest hundred

**63.** $5,311,192,809,000 5 trillion, 311 billion, 192 million, 809 thousand

**65.** 3 hours and 10 minutes is approximately 3 hours to the nearest hour.

**67.** 4 hours and 35 minutes is approximately 5 hours to the nearest hour.

**69.** 123 feet 5 inches is approximately 123 feet to the nearest foot.

**1.2 Understanding the Concept**

**1.** $8+5=(3+5)+5=3+(5+5)=3+10=13$

**2.** $6+8=6+(6+2)=(6+6)+2=12+2=14$

**1.2 Exercises**

**1.** $10+x$: ten plus a number

**3.** To evaluate $x+6$ if $x$ is equal to 9, replace $x$ with 9 and then add 9 and 6.

**5.** $(2+3)+4=2+(3+4)$ represents the associative property of addition.

**7.** A number plus 2: $m+2$

**9.** The sum of 5 and $y$: $5+y$

**11.** A number added to three: $m+3$

**13.** A number increased by 7: $m+7$

**15.** $2+x=x+2$

**17.** $8+n=n+8$

**19.** $216+3542=3758$

**21.** $n+5=12$

**23.** $n+7+3=n+10$

**25.** $6+2+x=8+x=x+8$

**27.** $x+0+2=x+2$

**29.** $(x+2)+1=x+(2+1)=x+3$

**31.** $5+(3+n)=(5+3)+n=8+n=n+8$

**33.** $(n+3)+8=n+(3+8)=n+11$

**35.** $(x+2)+8=x+(2+8)=x+10$

**37.** $(6+n)+3=(n+6)+3=n+(6+3)$
$=n+9$

**39.** $8+(1+x)=(8+1)+x$
$=9+x$
$=x+9$

**41.** $(3+n)+6=(n+3)+6$
$=n+(3+6)$
$=n+9$

**43.** $(7+a+1)+3=(a+7+1)+3$
$=(a+8)+3$
$=a+(8+3)$
$=a+11$

**45.** $(1+x+7)+2=(x+1+7)+2$
$=(x+8)+2$
$=x+(8+2)$
$=x+10$

**47.** (a) $y+7\big|_{y=3}=3+7=10$
(b) $y+7\big|_{y=8}=8+7=15$

**49.** $x+y\big|_{x=2,y=10}=2+10=12$

**51.** $a+b+c\big|_{a=9,b=15,c=12}=9+15+12=36$

**53.** $n+m+13\big|_{n=26,m=44}=26+44+13=83$

**55.** (a) $B=x+y+250$
$=180+12+250=\$442$
(b) $B=x+y+250$
$=175+10+250=\$435$

**57.**
$$\begin{array}{r} 36 \\ +\,23 \\ \hline 59 \end{array}$$

**59.**
$$\begin{array}{r} 142 \\ +\,34 \\ \hline 176 \end{array}$$

**61.**
$$\begin{array}{r} 21 \\ 14 \\ 8 \\ +\,7 \\ \hline 50 \end{array}$$

**63.**
$$\begin{array}{r} 105 \\ 8 \\ 133 \\ +\,98 \\ \hline 344 \end{array}$$

**65.** $236+467+26=729$

**67.** $397+29+467=893$

**69.** $7287+273+522=8082$

**71.** $922{,}876+54+1287+5000=929{,}217$

**73.** $3121+8050+16+163{,}942=175{,}129$

**75.** (a) $159+241=\$400$
(b) $63+121+44=\$228$

**77.** $2(875)+500+24+35=\$2309$

**79.** $P=2(11)+2(6)=34$ in.

**81.** $P=4(4)=16$ ft

**83.** $P=2(3)+5=11$ in.

**85.** $P=12+4+6+11+(12+6)+7=58$ ft

**87.** $P=40+30+135+(30+40)+175+40$
$P=490$ in.

**89.**
$P=120+100+190+(100+150)+310+150$
$P=1120$ in.

**91.** $(2+3)+4=2+(3+4)=2+7$ represents the associative property of addition.

**93.** $1+(x+3)+6=1+6+(x+3)=7+(x+3)$ represents the commutative property of addition.

### Cumulative Review

**95.** In the number 20,891 the digit 8 is in the hundreds place.

**97.** $102,876 = 103,000$ to nearest thousand

**1.3 Understanding the Concept**

**1.** We can borrow only from a place value that has a nonzero whole number. In $400 there are only 100-dollar bills to borrow from.

**2.** When we change the ten-dollar bill to 10 one-dollar bills, we have 0 ten-dollar bills and 10 one-dollar bills which is similar to borrowing in subtraction.

**1.3 Exercises**

**1.** $6 - x$: six minus $x$

**3.** The key phrase "how many more" indicates the operation subtraction.

**5.** $6 - 2 = 4$

**7.** $9 - 5 = 4$

**9.** $9 - 7 = 2$

**11.** $8 - 7 = 1$

**13.** $18 - 0 = 18$

**15.** $20 - 20 = 0$

**17.** $600 - 500 = 100$
$600 - 501 = 99$
$600 - 502 = 98$
$600 - 503 = 97$
$600 - 504 = 96$

**19.** $300 - 200 = 100$
$300 - 201 = 99$
$300 - 202 = 98$
$300 - 203 = 97$
$300 - 204 = 96$
$300 - 205 = 95$

**21.** Seven decreased by three: $7 - 3$

**23.** The difference of 8 and $y$: $8 - y$

**25.** Ten subtracted from seventeen: $17 - 10$

**27.** Fifteen minus a number: $15 - x$

**29.** Two less than some number: $a - 2$

**31.** $9 - n\big|_{n=0} = 9 - 0 = 9$

**33.** $9 - n\big|_{n=8} = 9 - 8 = 1$

**35.** $x - 2\big|_{x=3} = 3 - 2 = 1$

**37.** $x - 2\big|_{x=7} = 7 - 2 = 5$

**39.** $87 - 46 = 41$, check: $46 + 41 = 87$

**41.** $69 - 34 = 35$, check: $34 + 35 = 69$

**43.** $83 - 67 = 16$, check: $67 + 16 = 83$

**45.** $51 - 27 = 24$, check: $27 + 24 = 51$

**47.** $966 - 177 = 789$, check: $177 + 789 = 966$

**49.** $500 - 43 = 457$, check: $43 + 457 = 500$

**51.** $9922 - 2667 = 7255$
check: $2667 + 7255 = 9922$

**53.** $5301 - 185 = 5116$
check: $185 + 5116 = 5301$

**55.**
$$\begin{array}{r} 17,008 \\ -\ 4,839 \\ \hline 12,169 \end{array} \qquad \text{check:} \qquad \begin{array}{r} 4,839 \\ +\ 12,169 \\ \hline 17,008 \end{array}$$

**57.** $\begin{array}{r}164,300\\-58,923\\\hline 105,377\end{array}$ check: $\begin{array}{r}58,923\\+105,377\\\hline 164,300\end{array}$

**59.** $P = 4+3+(11-4)+8+11+(8-3)$
$P = 38$ ft

**61.** $P = 6+9+17+25+(25-9)+(17-6)$
$P = 84$ ft

**63.**

| Check # | Amount | Balance $1364 |
|---|---|---|
| 123 | $238 | $1126 |
| 124 | $137 | $ 989 |
| 125 | $ 69 | $ 920 |
| 126 | $ 98 | $ 822 |
| 127 | $369 | $ 453 |

**65.** $865,000 - 2160 = 862,840$ mi

**67.** $100 - 85 = 15$ mph faster

**69.** $300 - 195 = 105$ ft less

**71.** No, because the fifty-dollar bill cannot be converted into all twenties.

**73.** $8 - y = 8 - 3 = 5$

**Cumulative Review**

**75.** $5,117,206 > 13,842$

**75.** $31,007 + 579 = 31,586$

**77.** $120 + 135 + 105 = 360$ hours

**1.4 Understanding the Concept**

**1.** (a) $3(7) = 2(7) + 7 = 14 + 7 = 21$
(b) $4(8) = 5(8) - 8 = 40 - 8 = 32$
(c) $6(8) = 5(8) + 8 = 40 + 8 = 48$
(d) $9(8) = 10(8) - 8 = 80 - 8 = 72$

**1.4 Exercises**

**1.** (a) $4x$: Four times a number
(b) $ab$: The product of $a$ and $b$

**3.** Two times three:

```
            *  *
*  *  *
            *  *
*  *  *
            *  *
```

**5.** $3(6 \cdot 5) = (3 \cdot 6)5$ represents the associative property of multiplication.

**7.** $3 \cdot 4(2y) = 3 \cdot 4 \cdot \boxed{2} \cdot y = \boxed{24}y$

**9.** $(3a) \cdot 4 \cdot 2 = 3 \cdot \boxed{a} \cdot 4 \cdot 2 = 3 \cdot 4 \cdot 2 \cdot \boxed{a} = \boxed{24}a$

**11.** (a)

| | White | Pink | Blue |
|---|---|---|---|
| Brown | Brown White | Brown Pink | Brown Blue |
| Black | Black White | Black Pink | Black Blue |
| Gray | Gray White | Gray Pink | Gray Blue |
| Dark Blue | Dark Blue White | Dark Blue Pink | Dark Blue Blue |

(b) $4(3) = 12$ different outfits

**13.** $8(5) = 40$ different dishes

**15.** $6(3)=18$, 6,3: factors; 18: product

**17.** $22x=88$, $22,x$: factors; 88: product

**19.** A number times 7: $x(7)=7x$

**21.** The product of two different numbers: $xy$

**23.** Six times a number: $6x$

**25.** If $x \cdot y=0$ and $x=6$, then $y=0$.

**27.** If $x(y \cdot z)=40$, then $(x \cdot y) \cdot z=40$

**29.** $(3)(6)(2)(5)=18(10)=180$

**31.** $(7)(3)(2)(5)=21(10)=210$

**33.** $2 \cdot 4 \cdot 5 \cdot 0=0$

**35.** $2 \cdot 2 \cdot 3 \cdot 5=4 \cdot 15=60$

**37.** $9(6c)=54c$

**39.** $5(z \cdot 8)=40z$

**41.** $8(a \cdot 7)=56a$

**43.** $3(9 \cdot c)=27c$

**45.** $3(1)(x \cdot 8)=24x$

**47.** $9(2)(0 \cdot y)=0$

**49.** $6(3)(1 \cdot b)=18b$

**51.** $2 \cdot 3(5y)=30y$

**53.** $(6x)3 \cdot 7=(6x) \cdot 21=126x$

**55.** $3 \cdot 6(8y)=18(8y)=144y$

**57.** $9(637)=5733$

**59.** $7(602)=4214$

**61.** $398(300)=119,400$

**63.** $793(600)=475,800$

**65.**
$$\begin{array}{r} 76 \\ \underline{\times 68} \\ 608 \\ \underline{456\phantom{0}} \\ 5168 \end{array}$$

**67.**
$$\begin{array}{r} 99 \\ \underline{\times 94} \\ 396 \\ \underline{891\phantom{0}} \\ 9306 \end{array}$$

**69.**
$$\begin{array}{r} 847 \\ \underline{\times 56} \\ 5082 \\ \underline{4235\phantom{0}} \\ 47,432 \end{array}$$

**71.**
$$\begin{array}{r} 455 \\ \underline{\times 86} \\ 2730 \\ \underline{3640\phantom{0}} \\ 39,130 \end{array}$$

**73.** $762(309)=235,458$

**75.** $409(432)=176,688$

**77.** $8324(922) = 7,674,728$

**79.** $2009(651) = 1,307,859$

**81.** $12,107(808) = 9,782,456$

**83.** $61,711(1000) = 61,711,000$

**85.** $8(40) = \$320$

**87.** $15(25) = 375$ trees

**89.** $327(12) = \$3924$

**91.** $15(60) = 900$ rooms
$50(20) = 1000$ curtains, yes

**93.** (a) 20 °F (b) $4(19) = 76$ °F

**95.** $3(6a)(7b)2 = 18a(14b) = 252ab$

**97.** $4(3a)(2b)2 = 12a(4b) = 48ab$

**99.** (a)-(d) see table (e) 36 (f) There are only a few blank spaces left in the table, so there are not many multiplication facts to learn.

| | 0 | 1 | 2 | 3 | 4 | 5 | 6 | 7 | 8 | 9 |
|---|---|---|---|---|---|---|---|---|---|---|
| 0 | 0 | 0 | 0 | 0 | 0 | 0 | 0 | 0 | 0 | 0 |
| 1 | 0 | 1 | 2 | 3 | 4 | 5 | 6 | 7 | 8 | 9 |
| 2 | 0 | 2 | 4 | 6 | 8 | 10 | 12 | 14 | 16 | 18 |
| 3 | 0 | 3 | 6 | | | 15 | | | | |
| 4 | 0 | 4 | 8 | | | 20 | | | | |
| 5 | 0 | 5 | 10 | 15 | 20 | 25 | 30 | 35 | 40 | 45 |
| 6 | 0 | 6 | 12 | | | 30 | | | | |
| 7 | 0 | 7 | 14 | | | 35 | | | | |
| 8 | 0 | 8 | 16 | | | 40 | | | | |
| 9 | 0 | 9 | 18 | | | 45 | | | | |

### Cumulative Review

**101.** $7000 - 142 = 6858$

**103.** $168,406,000 = 168,410,000$

**105.** $920 - 455 = 465$ mi

### 1.5 Understanding the Concept

**1.** $a \div b = b \div a$ when $a$ and $b$ are equal

### 1.5 Exercises

**1.** $220 \div 4$

**3.** $225 \div n$

**5.** (b) and (c) are correct

**7.** Twenty-seven divided by a number:
$27 \div x$

**9.** Forty-two dollars divided equally among six people: $42 \div 6$

**11.** The quotient of thirty-six and three:
$36 \div 3$

**13.** The quotient of three and thirty-six:
$3 \div 36$

**15.** $36 \div 36 = 1$

**17.** $0 \div 66 = 0$

**19.** $14 \div 0$ undefined

**21.** $50 \div 6 = 8$ R2

**23.** $2597 \div 7 = 371$

$$
\begin{array}{l}
\phantom{7)}\,371 \\
7\overline{)2597} \\
\phantom{7)}\,\underline{21} \\
\phantom{7)2}\,49 \\
\phantom{7)2}\,\underline{49} \\
\phantom{7)29}\,7 \\
\phantom{7)29}\,\underline{7} \\
\phantom{7)29}\,0
\end{array}
$$

check: $7(371) = 2597$

**25.** $2097 \div 4 = 524 \text{ R1}$

$$
\begin{array}{l}
\phantom{4)}\,524 \\
4\overline{)2097} \\
\phantom{4)}\,\underline{20} \\
\phantom{4)20}\,9 \\
\phantom{4)20}\,\underline{8} \\
\phantom{4)2}\,17 \\
\phantom{4)2}\,\underline{16} \\
\phantom{4)20}\,1
\end{array}
$$

check: $4(524) + 1 = 2097$

**27.** $1268 \div 30 = 42 \text{ R8}$

$$
\begin{array}{l}
\phantom{30)1}\,42 \\
30\overline{)1268} \\
\phantom{30)}\,\underline{120} \\
\phantom{30)1}\,68 \\
\phantom{30)1}\,\underline{60} \\
\phantom{30)12}\,8
\end{array}
$$

check: $30(42) + 8 = 1268$

**29.** $632 \div 30 = 21 \text{ R2}$

$$
\begin{array}{l}
\phantom{30)6}\,21 \\
30\overline{)632} \\
\phantom{30)}\,\underline{60} \\
\phantom{30)6}\,32 \\
\phantom{30)6}\,\underline{30} \\
\phantom{30)63}\,2
\end{array}
$$

check: $30(21) + 2 = 632$

**31.** $5817 \div 19 = 306 \text{ R3}$

$$
\begin{array}{l}
\phantom{19)5}\,306 \\
19\overline{)5817} \\
\phantom{19)}\,\underline{57} \\
\phantom{19)5}\,117 \\
\phantom{19)5}\,\underline{114} \\
\phantom{19)581}\,3
\end{array}
$$

check: $19(306) + 3 = 5817$

**33.** $2093 \div 41 = 51 \text{ R2}$

$$
\begin{array}{l}
\phantom{41)20}\,51 \\
41\overline{)2093} \\
\phantom{41)}\,\underline{205} \\
\phantom{41)20}\,43 \\
\phantom{41)20}\,\underline{41} \\
\phantom{41)209}\,2
\end{array}
$$

check: $41(51) + 2 = 2093$

**35.** $1369 \div 19 = 72 \text{ R1}$

$$
\begin{array}{l}
\phantom{19)13}\,72 \\
19\overline{)1369} \\
\phantom{19)}\,\underline{133} \\
\phantom{19)13}\,39 \\
\phantom{19)13}\,\underline{38} \\
\phantom{19)136}\,1
\end{array}
$$

check: $19(72) + 1 = 1369$

**37.** $18{,}985 \div 27 = 703 \text{ R4}$

$$
\begin{array}{l}
\phantom{27)18}\,703 \\
27\overline{)18985} \\
\phantom{27)}\,\underline{189} \\
\phantom{27)189}\,85 \\
\phantom{27)189}\,\underline{81} \\
\phantom{27)1898}\,4
\end{array}
$$

check: $27(703) + 4 = 18{,}985$

**39.** $11,571 \div 34 = 340 \text{ R}11$

$$\begin{array}{r} 340 \\ 34\overline{)11571} \\ \underline{102}\phantom{00} \\ 137\phantom{0} \\ \underline{136}\phantom{0} \\ 11 \end{array}$$

check: $34(340) + 11 = 11,571$

**41.** $113,317 \div 223 = 508 \text{ R}33$

$$\begin{array}{r} 508 \\ 223\overline{)113317} \\ \underline{1115}\phantom{00} \\ 1817 \\ \underline{1784} \\ 33 \end{array}$$

check: $223(508) + 33 = 113,317$

**43.** $70,141 \div 36 = 515 \text{ R}101$

$$\begin{array}{r} 515 \\ 136\overline{)70141} \\ \underline{680}\phantom{00} \\ 214\phantom{0} \\ \underline{136}\phantom{0} \\ 781 \\ \underline{680} \\ 101 \end{array}$$

check: $136(515) + 101 = 70,141$

**45.** $60 \div 14 = 4 \text{ R}4$
4 tickets were donated to the PTA.

**47.** $85 \div 5 = \$17$ each

**49.** $1050 \div 6 = \$175$ per day

**51.** $124 \div 2 = 62$ ft

**53.** (a) $36 \div 9 = 4$ babies
(b) $3600 \div 9 = 400$ babies
(c) $3600 \div 17 = 211 \text{ R}13$
212 people

**55.** (a) $(32 \div 4) \div 2 = 8 \div 2 = 4$
(b) $32 \div (4 \div 2) = 32 \div 2 = 16$
(c) Division is not associative.

**Cumulative Review**

**57.** Seven plus $x$ equals eleven: $7 + x = 11$

**59.** $4031(202) = 814,262$

**61.** $1389 - 430 - 495 = 464$ miles

**1.6 Exercises**

**1.** What number squared is equal to 16?

**3.** $2 \cdot 2 = 2^2$

**5.** $a \cdot a \cdot a \cdot a \cdot a = a^5$

**7.** $8 = 8^1$

**9.** $3 \cdot 3 \cdot 3 \cdot 3 = 3^4$

**11.** $5 \cdot 5 \cdot a \cdot a \cdot a = 5^2 a^3$

**13.** $3 \cdot 3 \cdot z \cdot z \cdot z \cdot z \cdot z = 3^2 z^5$

**15.** $7 \cdot 7 \cdot 7 \cdot y \cdot y \cdot x \cdot x = 7^3 y^2 x^2$

**17.** $n \cdot n \cdot n \cdot n \cdot n \cdot 9 \cdot 9 = n^5 9^2$

**19.** (a) $y^3 = y \cdot y \cdot y$
(b) $7^5 = 7 \cdot 7 \cdot 7 \cdot 7 \cdot 7$

**21.** $2^3 = 2 \cdot 2 \cdot 2 = 8$

**23.** $5^2 = 5 \cdot 5 = 25$

**25.** $1^6 = 1 \cdot 1 \cdot 1 \cdot 1 \cdot 1 \cdot 1 = 1$

**27.** $7^2 = 7 \cdot 7 = 49$

**29.** $4^4 = 4 \cdot 4 \cdot 4 \cdot 4 = 256$

**31.** $10^1 = 10$

**33.** $5^3 = 5 \cdot 5 \cdot 5 = 125$

**35.** $10^6 = 10 \cdot 10 \cdot 10 \cdot 10 \cdot 10 \cdot 10 = 1{,}000{,}000$

**37.** $x^2 = (5)^2 = 25$

**39.** $a^4\big|_{a=1} = 1^4 = 1$

**41.** Seven to the third power: $7^3$

**43.** Nine squared: $9^2$

**45.** $2 \cdot 4 - 1 = 8 - 1 = 7$

**47.** $7^2 + 5 - 3 = 49 + 2 = 51$

**49.** $5 \cdot 3^2 = 5 \cdot 9 = 45$

**51.** $2 \cdot 2^2 = 2 \cdot 4 = 8$

**53.** $5^2 - 7 + 3 = 25 - 4 = 21$

**55.** $9 + 2 \cdot 2 = 9 + 4 = 13$

**57.** $9 + (6 + 2^2) = 9 + (6 + 4) = 9 + 10 = 19$

**59.**
$$\begin{aligned} 40 \div 5 \times 2 + 3^2 &= 40 \div 5 \times 2 + 9 \\ &= 8 \times 2 + 9 \\ &= 16 + 9 = 25 \end{aligned}$$

**61.** $2 \times 15 \div 5 + 10 = 30 \div 5 + 10 = 6 + 10 = 16$

**63.** $2^2 + 8 \div 4 = 4 + 2 = 6$

**65.** $\dfrac{(4 + 4 \div 2)}{(5 - 3)} = \dfrac{4 + 2}{2} = \dfrac{6}{2} = 3$

**67.** $\dfrac{(12 - 2)}{(25 \div 5 \times 2)} = \dfrac{10}{5 \times 2} = \dfrac{10}{10} = 1$

**69.**
$$\begin{aligned} 7 + 5(3 \cdot 4 + 7) - 2 &= 7 + 5(12 + 7) - 2 \\ &= 7 + 5(19) - 2 \\ &= 7 + 95 - 2 \\ &= 102 - 2 = 100 \end{aligned}$$

**71.**
$$\begin{aligned} 59 - 4(1 + 5 \cdot 2) + 4 &= 59 - 4(1 + 10) + 4 \\ &= 59 - 4(11) + 4 \\ &= 59 - 44 + 4 \\ &= 15 + 4 = 19 \end{aligned}$$

**73.**
$$\begin{aligned} 8 + 3(4 \cdot 5 + 9) - 11 &= 8 + 3(20 + 9) - 11 \\ &= 8 + 3(29) - 11 \\ &= 8 + 87 - 11 \\ &= 84 \end{aligned}$$

**75.**
$$\begin{aligned} &32 \cdot 6 - 4(4^3 - 5 \cdot 2^2) + 3 \\ &= 192 - 4(64 - 5 \cdot 4) + 3 \\ &= 192 - 4(64 - 20) + 3 \\ &= 192 - 4(44) + 3 \\ &= 192 - 176 + 3 = 16 + 3 = 19 \end{aligned}$$

**77.**
$$\begin{aligned} &12 \cdot 5 - 3\left(3^3 - 2 \cdot 3^2\right) + 1 \\ &= 12 \cdot 5 - 3\left(27 - 2 \cdot 9\right) + 1 \\ &= 60 - 3(27 - 18) + 1 \\ &= 60 - 3(9) + 1 \\ &= 60 - 27 + 1 \\ &= 34 \end{aligned}$$

**79.** He should have multiplied 3 times 2 first and then added 4 to get 10.

**81.** $21\cdot 10^1 = 210,\ 21\cdot 10^2 = 2100$
$21\cdot 10^3 = 21{,}000,\ 21\cdot 10^4 = 210{,}000$
The exponent on 10 determines the number of trailing zeros attached to the number.

**Cumulative Review**

**83.** $4079 + 2762 = 6841$

**85.** $(387)(196) = 75{,}852$

**87.** $15(140) = 2100 < 2200$, no.

**How Am I Doing in 1.1-1.6?**

**1.** $9062 = 9000 + 60 + 2$

**2.** $16 < 22$

**3.** $17{,}248{,}954 = 17{,}200{,}000$ to the nearest hundred thousand.

**4.** (a)
$$\begin{aligned}(6+a)+3 &= (a+6)+3\\ &= a+(6+3)\\ &= a+9\end{aligned}$$
(b)
$$\begin{aligned}(6+x+4)+2 &= (x+6+4)+2\\ &= (x+10)+2\\ &= x+(10+2)\\ &= x+12\end{aligned}$$

**5.** $x + y = 9 + 11 = 20$

**6.** $9532 + 251 + 322 = 10{,}105$

**7.** $P = 9 + 8 + (11-9) + 6 + 11 + (8+6)$
$P = 50$ in.

**8.** Eleven decreased by a number: $11 - x$

**9.** $39{,}204 - 5982 = 33{,}222$
check: $5982 + 33{,}222 = 39{,}204$

**10.** Double a number: $2x$

**11.** $2(4)(y\cdot 5) = 8(5\cdot y) = (8\cdot 5)y = 40y$

**12.** $(2371)126 = 298{,}746$

**13.** $6(12) = 72$ rooms

**14.** The quotient of 144 and $x$: $144 \div x$

**15.** $\dfrac{362{,}664}{721} = 503 \text{ R}1$

```
      503
721)362664
    3605
      2164
      2163
         1
```

**16.** $n\cdot n\cdot n\cdot n\cdot 3\cdot 3\cdot 3 = n^4\cdot 3^3 = 3^3n^4$

**17.** $4^3 = 4\cdot 4\cdot 4 = 64$

**18.** $2\cdot 3^2 = 2\cdot 9 = 18$

**19.**
$$\begin{aligned}(2+10)+12\div 6-3^2 &= (2+10)+12\div 6-9\\ &= 12+12\div 6-9\\ &= 12+2-9\\ &= 5\end{aligned}$$

**1.7 Exercises**

**1.** $5(3+4) = 5\cdot 3 + 5\cdot 4$ represents the distributive property of multiplication over addition

**3.** (a) $8(3y) = 8 \cdot 3 \cdot 8 \cdot y$ is false because the distributive property is only used when the terms inside the parentheses are being added or subtracted.
(b) $8(3+y) = 8 \cdot 3 + 8y$ is true because the terms inside the parentheses are being added.

**5.** $2(x+1) = 2 \cdot \boxed{x} + 2 \cdot \boxed{1}$

**7.** $6(y-3) = 6 \cdot \boxed{y} - 6 \cdot \boxed{3}$

**9.** six times $y$ plus two: $6y+2$

**11.** seven times four minus one: $7 \cdot 4 - 1$

**13.** four times the sum of three and nine: $4(3+9)$

**15.** six times the sum of $y$ and two: $6(y+2)$

**17.** eight times the difference of four and $y$: $8(4-y)$

**19.** (a) $4 \cdot 2 + 7 = 8 + 7 = 15$
(b) $4(2+7) = 4(9) = 36$

**21.** (a) $4 \cdot 3 - 1 = 12 - 1 = 11$
(b) $4(3-1) = 4(2) = 8$

**23.** (a) $12 \cdot 1 + 3 = 12 + 3 = 15$
(b) $12(1+3) = 12(4) = 48$

**25.** $4a + 5b = 4(2) + 5(6) = 8 + 30 = 38$

**27.** $8x - 6y = 8(7) - 6(2) = 56 - 12 = 44$

**29.** $\dfrac{x+4}{3} = \dfrac{11+4}{3} = \dfrac{15}{3} = 5$

**31.** $\dfrac{a^2-4}{b} = \dfrac{8^2-4}{6} = \dfrac{64-4}{6} = \dfrac{60}{6} = 10$

**33.** $\dfrac{x^3+4}{y} = \dfrac{2^3+4}{2} = \dfrac{8+4}{2} = \dfrac{12}{2} = 6$

**35.** $\dfrac{a^2+6}{b} = \dfrac{2^2+6}{5} = \dfrac{4+6}{5} = \dfrac{10}{5} = 2$

**37.** $\dfrac{y-2}{6} = \dfrac{20-2}{6} = \dfrac{18}{6} = 3$

**39.** $4m + 3n = 4(2) + 3(7) = 8 + 21 = 29$

**41.** $\dfrac{x^2-5}{y} = \dfrac{5^2-5}{4} = \dfrac{25-5}{4} = \dfrac{20}{4} = 5$

**43.** $4(x+2) = 4x + 8$

**45.** $3(n-5) = 3n - 15$

**47.** $3(x-6) = 3x - 18$

**49.** $4(x+4) = 4x + 16$

**51.** $3(x+1) + 5 = 3x + 3 + 5 = 3x + 8$

**53.** $2(y+6) + 2 = 2y + 12 + 2 = 2y + 14$

**55.** $4(x+1) + 6 = 4x + 4 + 6 = 4x + 10$

**57.** $2(y+6) - 3 = 2y + 12 - 3 = 2y + 9$

**59.** $2(x+2) - 1 = 2x + 4 - 1 = 2x + 3$

**61.** $2(y+5) + 3 = 2y + 10 + 3 = 2y + 13$

**63.** $yx^2 - 3 = 6 \cdot 2^2 - 3 = 6 \cdot 4 - 3 = 24 - 2 = 21$

**65.** $\frac{(a^2-3)+2^3}{b} = \frac{(5^2-3)+2^3}{2}$
$= \frac{(25-3)+8}{2} = \frac{22+8}{2}$
$= \frac{30}{2} = 15$

**67.** $3xy+2x+4y = 3\cdot4\cdot2+2\cdot4+4\cdot2$
$= 24+8+8 = 40$

**69.** (a)
$(x+2)+(x+2)+(x+2)+(x+2) = 4x+8$
(b) $4(x+2) = 4x+4(2) = 4x+8$
(c) The answers are the same.

**Cumulative Review**

**71.** $8(2)(x\cdot4) = 16(4x) = 64x$

**73.** $x+6=8 \Rightarrow x = 8-6 = 2$

**1.8 Exercises**

**1.** $7x$: seven times $x$ or the product of seven and $x$

**3.** $8x = 40$: eight times what number equals 40

**5.** $2x+3y$ cannot be added because they are not like terms.

**7.** In $6x+5$, $6x$is called a variable term and 5 is called a constant term.

**9.** The numerical part of $8x$ is 8 and it is called the coefficient of the term.

**11.** $y = \underline{1}y$

**13.** $2x+3\boxed{x} = 5x$

**15.** $3xy+\boxed{4xy} = 7xy$

**17.** $3x+2xy+\boxed{4x} = 7x+2xy$

**19.** Two $x$'s: $2x$

**21.** $a+a+a = 3a$

**23.** In $5x+3y+2x+8m+6y$, $5x$ and $2x$ are like terms; $3y$ and $6y$ are like terms.

**25.** In $2mn+3y+4mn+2$, $2mn$ and $4mn$ are like terms.

**27.** $7x+x = 8x$

**29.** $9y-3y = 6y$

**31.** $3x+2x+6x = 11x$

**33.** $8x+4a+3x+2a = 11x+6a$

**35.** $6xy+4b+3xy = 9xy+4b$

**37.** $6xy+3x+9+9xy = 15xy+3x+9$

**39.** $10ab-4ab+9 = 6ab+9$

**41.** $9xy+4+3xy+9 = 12xy+13$

**43.** $P = 2(5x)+2(4x+7y)$
$P = 10x+8x+14y = 18x+14y$

**45.** $P = 2(8a+2b)+2(6a+5b)$
$P = 16a+4b+12a+10b = 28a+14b$

**47.** $P = 6y+5x+2y+x = 6x+8y$

**49.** Four plus what number equals sixteen:
$4+x = 16$

**51.** What number times three equals thirty-six: $x \cdot 3 = 3x = 36$

**53.** If a number is subtracted from forty-five the result is six: $45 - x = 6$

**55.** Twenty-five divided by what number is equal to five: $\frac{25}{n} = 5$

**57.** $J + 10 = 25$

**59.** $C - 50 = 1480$

**61.** $8 - x\big|_{x=4} = 8 - 4 = 4 \neq 3$, no

**63.** $x + 4\big|_{x=15} = 15 + 4 = 19$, yes

**65.** $x + 3 = 9,\ x = 6$

check: $6 + 3 \stackrel{?}{=} 9,\ 9 = 9$

**67.** $9 - n = 7,\ n = 2$

check: $9 - 2 \stackrel{?}{=} 7,\ 7 = 7$

**69.** $x - 6 = 0,\ x = 6$

check: $6 - 6 \stackrel{?}{=} 0,\ 0 = 0$

**71.** $1 + x = 13,\ x = 12$

check: $1 + 12 \stackrel{?}{=} 13,\ 13 = 13$

**73.** $25 - x = 20,\ x = 5$

check: $25 - 5 \stackrel{?}{=} 20,\ 20 = 20$

**75.** $8x = 16,\ x = 2$

check: $8(2) \stackrel{?}{=} 16,\ 16 = 16$

**77.** $4y = 12,\ y = 3$

check: $4(3) \stackrel{?}{=} 12,\ 12 = 12$

**79.** $8x = 56,\ x = 7$

check: $8(7) \stackrel{?}{=} 56,\ 56 = 56$

**81.** $\frac{15}{y} = 1,\ y = 15$

check: $\frac{15}{15} \stackrel{?}{=} 1,\ 1 = 1$

**83.** $\frac{14}{x} = 2,\ x = 7$

check: $\frac{14}{7} \stackrel{?}{=} 2,\ 2 = 2$

**85.** $(x + 1) + 3 = x + 4 = 7,\ x = 3$

check: $(3 + 1) + 3 \stackrel{?}{=} 7,\ 7 = 7$

**87.** $7 + (2 + y) = 10,\ 2 + y = 3,\ y = 1$

check: $7 + (2 + 1) \stackrel{?}{=} 10,\ 10 = 10$

**89.** $3 + (n + 5) = 10,\ (n + 5) = 7,\ n = 2$

check: $3 + (2 + 5) \stackrel{?}{=} 10,\ 10 = 10$

**91.** $3n + n = 12,\ 4n = 12,\ n = 3$

check: $3(3) + 3 \stackrel{?}{=} 12,\ 12 = 12$

**93.** $7x - 2x - x = 4,\ 4x = 4,\ x = 1$

check: $7(1) - 2(1) - 1 \stackrel{?}{=} 4,\ 4 = 4$

**95.** $5x = 20,\ x = 4$

check: $5(4) \stackrel{?}{=} 20,\ 20 = 20$

**97.** $16-x=1,\ x=15$

check: $16-15\stackrel{?}{=}1,\ 1=1$

**99.** $2+(5+a)=14,\ (5+a)=12,\ a=7$

check: $2+(5+7)\stackrel{?}{=}14,\ 14=14$

**101.** $8x-5x=21,\ 3x=21,\ x=7$

check: $8(7)-5(7)\stackrel{?}{=}21,\ 21=21$

**103.** Four plus what number equals eight?
(a) $4+x=8$
(b) $x=4$

**105.** Three times what number equals nine?
(a) $3x=9$
(b) $x=3$

**107.** Triple what number times 2 is equals twelve?
(a) $3x\cdot 2=12$
(b) $6x=12,\ x=2$

**109.** $6+(2x^2+5)+(7+3x^2)+x^2=6x^2+18$

**111.** (a) $2x+5x+5y=5x+5y$
(b) $(2x)(5y)=10xy$

**113.** (a) $5a+6y+2a=7a+6y$
(b) $(5a)(6y)=30ay$

**115.** (a) 30 mph
(b) $50-2x,\ x=25$ mph

**117.** $2(50)+x=170$
$100+x=170$
$x=70$ ft

**Cumulative Review**

**119.** d

**121.** a

**1.9 Exercises**

**1.** (a) $80+40+20+10=\$150$
(b) $81+36+22+14=\$153$
(c) yes

**3.** (a) $6600-6500=100$
(b) $6551-6539=12$
(c) est. $-$ exact $=100-12=88$

**5.** First two days: $600+400=1000$
Second two days: $400+400=800$
$1000-800=200$ mi more the first two days.

**7.** $2(13)+4(5)+7=\$53$

**9.** $\dfrac{7(475)+20(65)+2(650)}{5}=\$1185$

**11.** $20+25+80+150+80+25+20=400$ ft

**13.** $40(8)+12(12)=\$464$

**15.** $35(7)=245$ gal in one day
$35(7)(7)=1715$ gal in one week

**17.** (a) $68,542-14,372=\$54,170$
(b) $54,170\div 2=\$27,085$

**19.** $15(12)(250)=45,000$ lb

**21.** $2(7)(4)(6)=\$336<\$400$
Cheaper to pay each time he rides.

**23.** $\frac{2(450)+2(5)+200}{4} = \$300$

**25.** (a) $6000 \div 2 = \$3000$
(b) $(6000 \div 2) \div 3 = \$1000$

**27.** (a) $30 + 240 + 170 = 440$
$440 \div 50 = 8 \text{ R}40 \rightarrow 8(5) = 40$ pts
$40 + 25(240 > 200) \rightarrow 65$ pts total
(b) $65 \div 10 = 6 \text{ R}5 \rightarrow \$6$ in discounts

**29.** (a) $170 + 260 = 430$
$430 \div 50 = 8 \text{ R}30 \rightarrow 8(10) = 80$ pts
$80 + 50(260 > 200) \rightarrow 130$ pts
(b) $130 \div 25 = 5 \text{ R}5,\ 5 \cdot \$5 = \$25$ discount

**31.** (a) $440,000 + 140,000 + 130,000$
$+210,000 = \$920,000$
(b) $435,750 + 144,800 + 125,500$
$+213,125 = \$919,175$
(c) yes

**33.** $175(36) - 5000 = \$1300$

**Cumulative Review**

**35.** $215 \div 5 = 43$

**37.** $x - 22 = 36$
$x = 58$

**Putting Your Skills to Work, Problems for Individual Analysis**

**1.** Plan 1:
$4(450) + 13(40) + 13(80)$
$+13(75) = \$4335$
Plan 2:
$350(2) + 175(2) + 2(3)(40)$
$+12(75) + 17(80) = \$3550$

**1.** Plan3:
$22(20) + 21(75) + 21(80) = \$3695$

**2.** Plan 2 is cheapest.

**3.** $3550 + 4(65) = \$3810$
Plan 3 is now cheapest.

**Putting Your Skills to Work, Problems for Group Investigation and Study**

**1.** Plan 1 is still \$4335 since no additional vacation days are required. Plan 2 is now $3550 + 240 = \$3790$ since it requires one additional vacation day. Plan 3 is now $3695 + 5(240) = \$4895$ since it requires one additional vacation day.

**2.** Plan 2 is now cheapest.

**Chapter 1 Review Problems**

**1.** rectangle-a four-sided figure with adjoining sides that are perpendicular and opposite sides that are equal.

**2.** square-a rectangle with all sides equal.

**3.** right angle-an angle that measures $90°$.

**4.** triangle-a three-sided figure with three angles.

**5.** perimeter-the distance around an object

**6.** factors-the numbers or variables that we multiply.

**7.** term-a number, a variable, or a product of a number and one or more variables.

**8.** constant term-a term that has no variable.

**9.** coefficient-the number factor in a term.

**10.** like terms-terms with identical variable parts.

**11.** equation-two expressions separated by an equal sign.

**12.** (a) In the number 175,493 the digit 7 is in the ten thousands place.
(b) In the number 175,493 the digit 5 is in the thousands place.

**13.** (a) In the number 458,013 the digit 8 is in the thousands place.
(b) In the number 458,013 the digit 0 is in the hundreds place.

**14.** $7694 = 7000 + 600 + 90 + 4$

**15.** $4325 = 4000 + 300 + 20 + 5$

**16.** Three hundred forty-one and 00/100

**17.** One hundred eighty-seven and 00/100

**18.** $2 < 8$

**19.** $12 > 11$

**20.** $11 < 12$

**21.** Six is greater than one: $6 > 1$

**22.** Three is less than five: $3 < 5$

**23.** $61,269 = 61,300$ to nearest hundred

**24.** $382,240 = 382,200$ to nearest hundred

**25.** $6,365,534 = 6,400,000$ to nearest hundred thousand

**26.** $8,118,701 = 8,100,000$ to nearest hundred thousand

**27.** Seven more than $x$: $x + 7$

**28.** The sum of some number and five: $n + 5$

**29.** A number increased by four: $x + 4$

**30.** $7 + (9 + x) = (7 + 9) + x = 16 + x = x + 16$

**31.** $(n + 4) + 1 = n + (4 + 1)$
$= n + 5$

**32.** $(2 + n) + 9 = (n + 2) + 9$
$= n + (2 + 9)$
$= 11$

**33.** $5 + (n + 2) = (n + 2) + 5$
$= n + (2 + 5)$
$= n + 7$

**34.** $(5 + x + 3) + 2 = (x + 5 + 3) + 2 = x + 10$

**35.** $(3 + x + 4) + 1 = (x + 3 + 4) + 1 = x + 8$

**36.** $7 + n\big|_{n=6} = 7 + 6 = 13$

**37.** $x + 3\big|_{x=17} = 17 + 3 = 20$

**38.** $x + y\big|_{x=11, y=15} = 11 + 15 = 26$

**39.** $8398 + 372 + 255 = 9025$

**40.** $17,456 + 213 + 982 = 18,651$

**41.** $1434 + 1596 + 1423 + 1565 = 6018$

**42.** $13 + 5 + 7 + (5 + 8) + (13 + 7) + 8 = 66$ m

**43.** Eight decreased by 3: $8-3$

**44.** The difference of three and a number: $3-y$

**45.** Ten subtracted from a number: $x-10$

**46.** $8-x\big|_{x=3}=8-3=5$

**47.** $a-2=19-2=17$

**48.** $7-x\big|_{x=2}=7-2=5$

**49.** $8-y=8-3=5$

**50.** $8502-2957=5545$
check: $5545+2957=8502$

**51.** $9021-5862=3159$
check: $3159+5862=9021$

**52.** $29{,}104-4988=24{,}116$
check: $24{,}116+4988=29{,}104$

**53.** $137{,}405-6877=130{,}528$
check: $130{,}528+6877=137{,}405$

**54.** $900{,}000-522{,}000=\$378{,}000$ more

**55.** $720{,}000-450{,}000=\$270{,}000$ less

**56.** $4x=32$, factors: 4, $x$; product: 32

**57.** $xy=z$, factors: $x$, $y$; product: $z$

**58.** Triple a number: $3x$

**59.** $7y=63$, What number times 7 equals 63?

**60.** $7\cdot 2\cdot 3\cdot 0=0$

**61.** $5\cdot 3\cdot 2\cdot 2=15\cdot 4=60$

**62.** $2(y\cdot 7)=2(7y)=14y$

**63.** $3(7)(x\cdot 2)=21(2x)=42x$

**64.** $2(5)(y\cdot 7)=10(7y)=70y$

**65.** $3(2)(x\cdot 4)=24x$

**66.** $572(71)=40{,}612$

**67.** $(416)(2000)=832{,}000$

**68.** $(4251)352=1{,}496{,}352$

**69.** $6424\times 903=5{,}800{,}872$

**70.** $17(18)=306$ mi

**71.** $6(21)(4)=504$ doors

**72.** $300\div 20$

**73.** $\$500\div n$

**74.** Five divided by a number: $5\div y$

**75.** $26\div 13$

**76.** $10\div 0$ undefined

**77.** $33\div 33=1$

**78.** $4\overline{)1804}$ = 451

**79.** $7\overline{)1704}$ = 243

**80.** $2485 \div 31 = 80 \text{ R5}$

**81.** $1456 \div 29 = 50 \text{ R6}$

**82.** $369,757 \div 922 = 401 \text{ R35}$

**83.** $\frac{510,144}{846} = 603 \text{ R6}$

**84.** $447 \div 4 = 111 \text{ R3} \rightarrow \$3$ deposit

**85.** $3528 \div 24 = \$147$

**86.** $2 \cdot 2 \cdot 2 \cdot n \cdot n = 2^3 n^2$

**87.** $z \cdot z \cdot z \cdot z \cdot 5 \cdot 5 \cdot 5 = z^4 \cdot 5^3$

**88.** $a^4 = a \cdot a \cdot a \cdot a$

**89.** $6^5 = 6 \cdot 6 \cdot 6 \cdot 6 \cdot 6$

**90.** $10^3 = 10 \cdot 10 \cdot 10 = 1000$

**91.** $2^5 = 2 \cdot 2 \cdot 2 \cdot 2 \cdot 2 = 32$

**92.** $9^2 = 9 \cdot 9 = 81$

**93.** Four squared: $4^2$

**94.** $x$ to the fifth power: $x^5$

**95.** $6 + 14 \div 2 - 2^2 = 6 + 7 - 4 = 9$

**96.** $(15 + 25 \div 5) \div (8 - 4) = (15 + 5) \div 4$
$= 20 \div 4 = 5$

**97.** $5 \cdot 2^2 = 5 \cdot 4 = 20$

**98.** (a) $3x + 2$ (b) $3(x + 2)$

**99.** (a) $4x - 5$ (b) $4(x - 5)$

**100.** (a) $3 \cdot 7 + 1 = 22$ (b) $3(7 + 1) = 24$

**101.** $x^3 - 1 = 3^3 - 1 = 27 - 1 = 26$

**102.** $2m + 3n = 2(8) + 3(2) = 16 + 6 = 22$

**103.** $2(x + 1) = 2x + 2(1) = 2x + 2$

**104.** $4(x - 1) = 4x - 4(1) = 4x - 4$

**105.** $3(x + 1) + 5 = 3x + 3 + 5 = 3x + 8$

**106.** $4x + x + 6x = 11x$

**107.** $3x + 2y + 6x = 9x + 2y$

**108.** $3xy + 5y + 2xy + 8y = 5xy + 13y$

**109.** $P = 2(2x + 4y) + 2(3x + y) = 10x + 10y$

**110.** $x + 2 = 9 \Rightarrow x = 7$
check: $7 + 2 \stackrel{?}{=} 9,\ 9 = 9$

**111.** $8 - n = 2 \Rightarrow n = 6$
check: $8 - 6 \stackrel{?}{=} 2,\ 2 = 2$

**112.** $(3 + x) + 1 = 8,\ x + 4 = 8,\ x = 4$

**113.** $2 + (n + 7) = 10,\ n + 9 = 10,\ n = 1$

**114.** $9x = 27,\ x = 3$ check: $9(3) \stackrel{?}{=} 27,\ 27 = 27$

**115.** $\frac{20}{x} = 5,\ x = 4$ check: $\frac{20}{4} \stackrel{?}{=} 5,\ 5 = 5$

**116.** $12n - n = 22,\ 11n = 22,\ n = 2$

**117.** $y + 3y + 2y = 12,\ 6y = 12,\ y = 2$

**118.** (a) $18 - x = 3$ (b) $x = 15$

**119.** (a) $x + 5 = 11$ (b) $x = 6$

**120.** (a) $3(x \cdot 2) = 12,\ 6x = 12,$ (b) $x = 2$

**121.** $30 + 30 + 90 + 160 = \$310$

**122.** $3560 - 499 - 218 - 97 = \$2746$

**123.** (a) $5021 + 759 + 2534 + 532 - 799 - 533 - 88 = \$7426$
(b) $7426 \div 2 = \$3713$

**124.** $(2(20) + 2(25) + 2(15) + 2(18))(3) = \$468$

**How Am I Doing? Chapter 1 Test**

**1.** $1525 = 1000 + 500 + 20 + 5$

**2.** (a) $7 > 2$ (b) $5 > 0$

**3.** (a) $2925 = 3000$ to nearest thousand
(b) $2925 = 2900$ to nearest hundred

**4.** (a) $3 + (8 + x) = (x + 8) + 3 = x + 11$
(b) $5 + y + 2 = y + 5 + 2 = y + 7$
(c) $1 + (n + 2) + 4 = (n + 2) + 5 = n + 7$

**5.** $12{,}389 + 4 + 2302 = 14{,}695$

**6.** $244{,}869{,}201 + 19{,}077 = 244{,}888{,}278$

**7.** (a) $613 - 75 = 538$, check: $75 + 538 = 613$
(b) $20{,}105 - 7{,}826 = 12{,}279$
check: $7826 + 12{,}279 = 20{,}105$

**8.** $2 + 1 + 7 + 5 + (7 + 2) + 6 = 30$ ft

**9.** $2(4)(y \cdot 2) = 8(2y) = 16y$

**10.** (a) $(432)(312) = 134{,}784$
(b) $2031 \times 129 = 261{,}999$

**11.** (a) $492 \div 12 = 41$ ck: $12(41) = 492$
(b) $5523 \div 46 = 120$ R3 ck: $120(46) + 3 = 5523$

**12.** (a) $n - 7$
(b) $10n$
(c) $y^4$
(d) $7^3$
(e) $6(x + 9)$

**13.** (a) $3xy + 2y + 4xy - 2 = 7xy + 2y - 2$
(b) $2m + 5 + m + 6mn = 3m + 5 + 6mn$

**14.** $3(y + 4) = 3y + 3(4) = 3y + 12$

**15.** $8(x + 1) + 2(x + 2) = 10x + 12$

**16.** (a) $2x - 3y = 2(16) - 3(4) = 20$
(b) $a^2 - 4 = 9^2 - 4 = 81 - 4 = 77$

**17.** $6 \cdot 6 \cdot 6 \cdot 6 \cdot 6 = 6^5$

**18.** (a) $5^3 = 5 \cdot 5 \cdot 5 = 125$
(b) $10^5 = 10 \cdot 10 \cdot 10 \cdot 10 \cdot 10 = 100{,}000$

**19.** $24 \div 4 - 2 \cdot 3 = 6 - 6 = 0$

**20.** $6^2 - 7 + 3 \cdot 4 = 36 - 7 + 12 = 29 + 12 = 41$

**21.** $3 \cdot 2 + 4(7 - 1) = 6 + 4(6) = 6 + 24 = 30$

**22.** (a) $7 + x = 13 \Rightarrow x = 6$
check: $7 + 6 \stackrel{?}{=} 13,\ 13 = 13$
(b) $\frac{x}{4} = 2 \Rightarrow x = 8$
check: $\frac{8}{4} \stackrel{?}{=} 2,\ 2 = 2$

**22.** (c) $x+3x=36,\ 4x=36,\ x=9$

check: $9+3(9)\stackrel{?}{=}36,\ 36=36$

(d) $5+(b+2)=18 \Rightarrow (b+2)+5=18$

$b+(2+5)=18$

$b+7=18$

$b=11$

check: $5+(11+2)\stackrel{?}{=}18,\ 18=18$

(e) $9n-n=32 \Rightarrow 8n=32 \Rightarrow n=4$

check: $9(4)-4\stackrel{?}{=}32,\ 32=32$

**23.** $B-155=275$

**24.** (a) $x \div 6=2$ (b) $x=12$

**25.** (a) $x-3=1$ (b) $x=4$

**26.** (a) $25(412)+18(280)=\$15,340$

(b) $P=15,340-7350=\$7990$

**27.** $4(3)=12$ different sandwiches

**28.** $1540-265-78-57=\$1140$

**29.** $2(525)+200+40=\$1290$

**30.** (a) $800+200+100+200+300=\$1600$

(b) $1900-1600=\$300$

**31.** $(5000 \div 2) \cdot 3=7500$ points

# Chapter 2

## 2.1 Exercises

**1.** $-(-1)$: the opposite of negative one

**3.** Negative four minus two: $-4-2$

**5.** $-9$ is a <u>negative</u> number and 9 is a <u>positive</u> number.

**7.** The set of whole numbers and their opposites is called <u>integers</u>.

**9.** 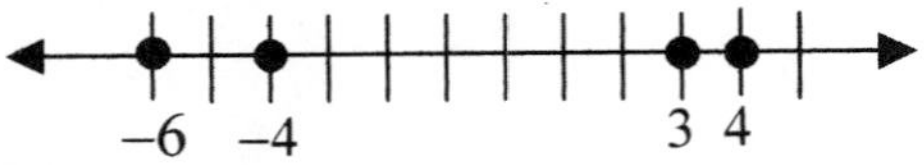


**11.** 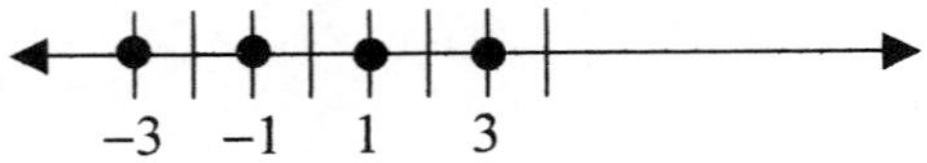


**13.** A represents the larger number.

**15.** $-5<-2$

**17.** $9>-3$

**19.** $-5<5$

**21.** $-7<-5$

**23.** $-12<-5$

**25.** $-298>-350$

**27.** ___+___ A tax increase of \$100

**29.** ___−___ A loss of \$100

**31.** ___−___ A discount of \$10

**33.** ___+___ A plane ascends 1000 ft

**35.** 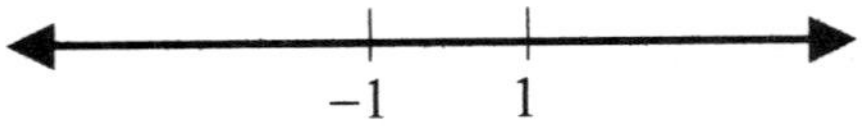


**37.** The opposite of $-5$ is ___5___.

**39.** The opposite of 16 is ___−16___.

**41.** $-(-11)=$ ___11___

**43.** $-(5)=$ ___−5___

**45.** $-(-(-7))=-(7)=-7$

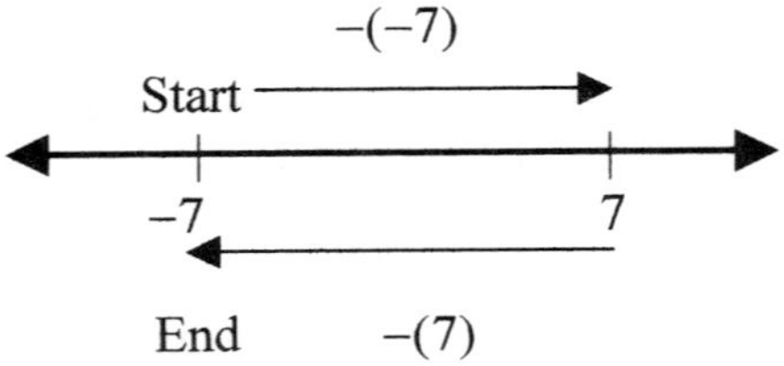


**47.** $-(-(13))=-(-13)$
$=13$

**49.** $-(-(-(-1))))=-(-(1))=1$

**51.** $-(-a)\big|_{a=6}=-(-6)$
$=6$

**53.** $-(-(-x))=-(x)=-x\big|_{x=-1}$
$=-(-1)=1$

**55.** $-(-(-(-y)))=-(-(y))=-(-y)$
$=y\big|_{y=-2}=-2$

**57.** $|8|=8$

**59.** $|-5|=-(-5)=5$

**61.** $|-16| = -(-16) = 16$

**63.** $|44| = 44$

**65.** $|-3| = 3,\ |1| = 1,\ 3 > 1 \Rightarrow |-3| > |1|$

**67.** $|8| = 8,\ |-8| = 8, \Rightarrow |8| = |-8|$

**69.** $|16| = 16,\ |-9| = 9 \Rightarrow |16| > |-9|$

**71.** $|-35| = 35,\ |-8| = 8 \Rightarrow |-35| > |-8|$

**73.** $-|-3| = -(-(-3)) = -(3) = -3$

**75.** $-|14| = -14$

**77.** (a) Fargo
(b) Positive: Boston and Cleveland
Negative: Albany, Anchorage, Fargo

**79.** $-|-16| = -16,\ -|-16| ? -16$ is
$-|-16| = -16$

**81.** $|-8| = 8,\ -|-8| = -8,\ |-8| ? -|-8|$ is $8 > -8$

**83.** $-(-6) + |-5| = 6 + 5 = 11$

**85.** $-(-|8|) - |-1| = -(-8) - 1 = 8 - 1 = 7$

**87.** $|-33| = 33,\ |12| = 12 \Rightarrow |-33| > |12|$
$-33$ has the larger absolute value.

**89.** $|129| = 129,\ |-112| = 112 \Rightarrow |129| > |-112|$
129 has the larger absolute value.

**91.** If $x > y$ then $x$ must be a positive number is false, for example $-1 > -2$.

**93.** If $m$ is a negative number then $-m$ is a positive number.

**Cumulative Review**

**95.** $5009 - 258 = 4751$

**97.** $(256)(91) = 23{,}296$

**99.** $2600 - 400 - 1200 - 350 = \$570$

**2.2 Exercises**

**1.** When two negative numbers are added the first move in the negative direction is followed by a second move in the negative direction resulting in a negative.

**3.** To add two numbers with different signs we keep the sign of the larger absolute value and subtract.

**5.** (a) $-2 + (-3) = \boxed{-5}$ Rule: When adding two numbers with the same sign, we use the common sign in the answer and add the absolute values of the numbers.
(b) $2 + (-3) = \boxed{-1}$ Rule: When adding two numbers with different signs we keep the sign of the larger absolute value and subtract the absolute values.
(c) $-2 + 3 = \boxed{+1}$ Rule: When adding two numbers with different signs we keep the sign of the larger absolute value and subtract the absolute values

**7.** (a)

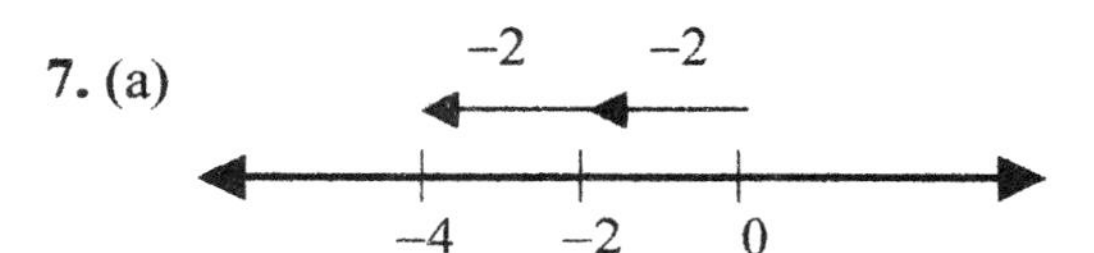

**7.** (b) Negative
(c) $-2+(-2)$
(d) From number line, sum is $-4$

**9.** (a)

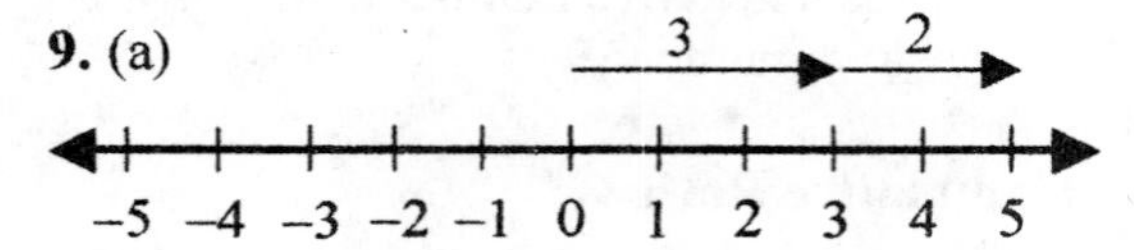


(b) Positive
(c) $3+2$
(d) From the number line, sum is 5.

**11.** (a) A decrease of 10°F followed by a decrease of 5°F results in a <u>decrease of 15°F</u>.
(b) $-10°F + -5°F = -15°F$

**13.** (a) A profit of \$100 followed by a profit of \$50 results in a <u>profit of \$150</u>.
(b) <u>\$100 + \$50 = \$150</u>

**15.** (a) $-11+(-13)=-24$
(b) $11+13=24$

**17.** (a) $-29+(-39)=-68$
(b) $29+39=68$

**19.** (a) $-43+(-18)=-61$
(b) $43+18=61$

**21.** (a) $-3+10$ (b) Positive
(c) $-3+10=7$

**23.** (a) $2+(-4)$ (b) Negative
(c) $2+(-4)=-2$

**25.** $+300 \text{ ft}+(-400 \text{ ft})=-100 \text{ ft}$

**27.** $-\$400+(+\$500)=\$100$

**29.** (a) $6+(-8)=-2$ (b) $-6+8=2$

**31.** (a) $5+(-1)=4$ (b) $-5+1=-4$

**33.** (a) $22+(-16)=6$ (b) $-22+16=-6$

**35.** (a) $3+(-1)=2$ (b) $-3+(-1)=-4$
(c) $-3+1=-2$

**37.** (a) $-9+(-11)=-20$
(b) $9+(-11)=-2$ (c) $-9+11=2$

**39.** $2+(-2)=0$

**41.** $-8+8=0$

**43.** $-350+350=0$

**45.** $452+(-452)=0$

**47.** $x+29=0,\ x=-29$

**49.** $-12+x=0,\ x=12$

**51.** $12+(-11)=1$

**53.** $-9+6=-3$

**55.** $-7+14=7$

**57.** $22+(-10)=12$

**59.** $-33+(-5)=-38$

**61.** $-27+(-12)=-39$

**63.** $-15+15=0$

**65.** $-12+16=4$

**67.** $6+(-8)=-2$

**69.** $14+(-14)=0$

**71.** $6+(-9)+1+(-3)=-3+(-2)=-5$

**73.** $-21+16+(-33)=-5+(-33)=-38$

**75.** $17+(-29)+(-34)+43=-12+9=-3$

**77.** $-15+7+(-10)+3=-8+(-10)+3$
$=-18+3=-15$

**79.** (a) $x+3\big|_{x=-2}=-2+3=1$
(b) $x+3\big|_{x=-7}=-7+3=-4$

**81.** (a) $x+(-8)\big|_{x=4}=4+(-8)=-4$
(b) $x+(-8)\big|_{x=-5}=-5+(-8)=-13$

**83.** (a) $-2+x+y\big|_{x=-6,y=4}=-2+(-6)+4=-4$
(b) $-2+x+y\big|_{x=5,y=-9}=-2+5+(-9)=-6$

**85.** $-x+y+6=-(-2)+(-1)+6=2+5=7$

**87.** $-a+b+(-1)=-(-3)+(-5)+(-1)$
$=3+(-6)=-3$

**89.** $30,000+(-20,000)=\$10,000$

**91.** $-121+200=\$79$

**93.** $-75+(-150)=-225$ ft

**95.** $3+x+y+-1)+z$
$=3+(-1)+9+(-1)+(-5)$
$=2+9+(-6)$
$=11+(-6)=5$

**97.** $-33+24+(-38)+19+(-3)$
$=-9+(-19)+(-3)=-28+(-3)=-31$

**99.** $12+(-45)+(-9)+5+(-19)$
$=-33+(-4)+(-19)$
$=-37+(-19)=-56$

**101.** $4568+(-3442)+9002+(-2,334,678)$
$=1126+(-2,325,676)=-2,324,550$

**103.** $-2+\boxed{-3}=-5$

**105.** $3+\boxed{-4}=-1$

**107.** 1 can be added to $-11$ to obtain $-11$.

**109.** If $x+y+30=0$, then $x+y=-30$. Two possible values for $x$ and $y$ are $x=-15,\ y=-15$.

**Cumulative Review**

**111.** $4x+6x=10x$

**113.** $8x-3x=5x$

**115.** $23,566+2(150+110)=24,086$ mi

**2.3 Understanding the Concept**

**1.** (a) $2-6-8-11=2+(-25)=-23$
(b)
$2-6-8-11=2+(-6)-8-11$
$=-4-8-11=-4+(-8)-11$
$=12-11=-12+(-11)=-23$

**2.3 Exercises**

**1.** To subtract two numbers, we change the subtraction sign to <u>addition</u> and take the <u>opposite</u> of the second number. Then we <u>add</u>.

**3.** We can think of this subtraction as measuring the distance between the two numbers. $-10$ is 10 units below 0 on a vertical number line, so we must add these 10 units to the 25 units above 0.

**5.** To subtract $-3$ we add $\boxed{3}$.

**7.** To subtract 5 we add $\boxed{-5}$.

**9.** $-7-6=-7+\boxed{-6}=\boxed{-13}$

**11.** $4-9=4+\boxed{-9}=\boxed{-5}$

**13.** $6-(-3)=6\boxed{+}3=\boxed{9}$

**15.** $9-(-5)=9\boxed{+}5=\boxed{14}$

**17.** (a) $7-4=7+(-4)=3$
(b) $15-7=15+(-7)=8$
(c) $10-8=10+(-8)=2$

**19.** $\$20-\$30=\$20+(-\$30)=-\$10$

**21.** $\$7-\$8=\$7+(-\$8)=-\$1$

**23.** $-6-4=-6+(-4)=-10$

**25.** $-5-4=-5+(-4)=-9$

**27.** $4-(-2)=4+(-(-2))=4+2=6$

**29.** $5-(-9)=5+(-(-9))=5+9=14$

**31.** $-8-(-6)=-8+(-(-6))=-8+6=-2$

**33.** $-8-(-8)=-8+(-(-8))=-8+8=0$

**35.** $2-7=2+(-7)=-5$

**37.** $3-7=3+(-7)=-4$

**39.** $80-90=80+(-90)=-10$

**41.** $-85-(-20)=-85+(-(-20))$
$=-85+20=-65$

**43.** $7-9-3-8=-2-3-8=-5-8=-13$

**45.** $2-1-9-7=1-9-7=-8-7=-15$

**47.** $9-10-2+3=-1-2+3=-3+3=0$

**49.** $-6-(-3)+(-7)=-3+(-7)=-10$

**51.** $-7-(-2)-(-5)=-5-(-5)=0$

**53.** $-3-(-8)+(-6)=5+(-6)=-1$

**55.** $5-19=5+(-19)=-14$

**57.** $-9-(-9)=-9+(9)=0$

**59.** $-13-18=-13+(-18)=-31$

**61.** $52-(-1)=52+(-(-1))=52+1=53$

**63.** $8-1-9-5=7-9-5=-2-5=-7$

**65.** $5+8-6-4=13-6-4=7-4=3$

**67.** $9-10-2+3=-1+1=0$

**69.** $x-12=-8-12=-20$

**71.** $x-11=-3-11=-14$

**73.** $14-y=14-(-5)=14+5=19$

**75.** $21-y+x=21-(-1)+2=22+2=24$

**77.** $-8-x-y=-8-(-4)-2=-4-2=-6$

**79.** $-1-x+y=-1-(-6)+(-1)$
$=5+(-1)=4$

**81.** $3556-(-150)=3706$ ft

**83.** $326-(-18)=344$ ft

**85.** (a) Brownsville, Texas
(b) $77\ °F-(-9\ °F)=86\ °F$

**87.** $-11-(-15)=4$ points

**89.** $-15<-13 \rightarrow$ Woods is the winner.

**91.** $-31+24-13-12+(-14)-3$
$=-7-25+(-17)=-32+(-17)=-49$

**93.** $7-a-b+c+2$
$=7-(-13)-(-4)+(-9)+2$
$=20-5+2=15+1=17$

**95.** $-3009-893=-3000+(-893)$
$=-3902$

**97.** $-2001-(-987)=-2001+(-(-987))$
$=-2001+987=-1014$

**99.** (a) $-1-8=n$ (b) $n=-9$

**101.** $-4$ is a solution to $x-3=-7$ since $-4-3=-7$.

**103.** $-2-\boxed{2}=-4$

**Cumulative Review**

**105.** $12-3(4-1)=12-3(3)=12-9=3$

**107.** $3+[3+2(8-6)]=3+[3+2(2)]$
$=3+[3+4]=3+7=10$

**109.** $8670 \div 85=102$ min $=1$ hr $42$ min

**Putting Your Skills to Work, Problems for Individual Analysis**

**1.** $|0-(-3)|=3$ hr earlier
10 A.M. − 3 hr gives 7 A.M. in Montevideo, Uruguay.

**2.** $|0-9|=9$ hr later
2 P.M. + 9 hr gives 11 P.M. in Seoul, South Korea.

**3.** $|3-(-7)|=10$ hr earlier
3 P.M. − 10 hr gives 5 A.M. in Denver, Colorado.

**4.** $|8-5|=3$ hr earlier
6 A.M. − 3 hr gives 3 A.M. in Karachi, Pakistan.

**Putting Your Skills to Work, Problems for Group Investigation and Study**

**1.** $|10-(-6)|=16$ hr earlier
9 A.M. Monday − 16 hr gives 5 P.M on Sunday in Mexico City, Mexico.

**2.** 2 P.M. Tuesday − 16 hr gives 10 P.M. on Monday in Mexico City, so, yes, Victor can watch since he got home at 6 P.M.

**3.** Since the call takes 30 min it must be made by 5 P.M. in Bucharest which is noon in Montevideo and 10 A.M. in Boston. Since the earliest the call can be made is 8 A.M. in Boston which is 10 A.M. in Montevideo and 1 P.M. in Bucharest, the call must be made between 8 A.M. and 10 A.M. from the Boston office.

**4.** Since the call takes 30 min it must be made by 4 P.M. in Kurachi which is noon in Algiers and 8 A.M. in Montvideo. The engineer in Algiers must make the call at noon.

### How Am I Doing? Sections 2.1-2.3

**1.** $-12 < -7$

**2.** $|-11| = 11,\ |8| = 8,\ |-11| = 11 > |8| = 8$

**3.** $-|-8| = -8$

**4.** $-(-(-3))) = -(3) = -3$

**5.** $-(-x) = -(-(-6)) = -(6) = -6$

**6.** $-1 + (-14) = -16$

**7.** $-8 + 3 + (-1) + 4 = -5 + 3 = -2$

**8.** $a + b + 12 = -9 + (-5) + 12$
$= -14 + 12 = -2$

**9.** $-x + y + 7 = -(-8) + (-11) + 7$
$= 8 + (-4) = 4$

**10.** $-20 + 20 + (-10) + 30 = 0 + 20 = 20$
The overall profit was $20,000

**11.** $7 - 19 = 7 + (-19) = -12$

**12.** $-3 - (-5) = -3 + 5 = 2$

**13.** $-8 - (-2) - (-1) = -8 + 2 + 1 = -5$

**14.** $-5 - 6 + (-1) - (-7) = -5 + (-6) + (-1) + 7$
$= -11 + 6 = -5$

**15.** $-5 - x - y = -5 - (-1) - (-2)$
$= -5 + 1 + 2 = -2$

**16.** $7622 - (-161) = 7783$ ft

### 2.4 Exercises

**1.** Adding two negative numbers gives a negative sum. Multiplying two negative numbers gives a positive product. Two negatives make a positive is true only for multiplication.

**3.** If you multiply 4 negative numbers, the product will be a <u>positive</u> number.

**5.** The quotient of a positive number and a <u>negative</u> number is negative.

**7.** $3(-4) = (-4) + (-4) + (-4) = -12$

**9.** $4(-6) = (-6) + (-6) + (-6) + (-6) = -24$

**11.** $2(-3) = (-3) + (-3) = -6$

**13.** (a) $3 \cdot \boxed{3} = 9$　(b) $3 \cdot \boxed{-3} = -9$
(c) $-3 \cdot \boxed{3} = -9$　(d) $-3 \cdot \boxed{-3} = 9$

**15.** (a) $\frac{12}{\boxed{4}} = 3$　(b) $\frac{-12}{\boxed{-4}} = 3$
(c) $\frac{12}{\boxed{-4}} = -3$　(d) $\frac{-12}{\boxed{4}} = -3$

**17.** (a) $4(2) = 8$　(b) $4(-2) = -8$
(c) $-4(2) = -8$　(d) $-4(-2) = 8$

**19.** (a) $5(2)=10$ (b) $-5(-2)=10$
(c) $-5(2)=-10$ (d) $5(-2)=-10$

**21.** $(-2)(-9)=18$

**23.** $-3(-6)=18$

**25.** $5(-6)=-30$

**27.** $-8(3)=-24$

**29.** $(-1)(3)(-236)(42)(-16)(-90)$ is a positive product because it contains an even number of negative factors.

**31.** $(-943)(-721)(-816)(-96)(-51)$ is a negative product because it contains an odd number of negative factors.

**33.** $3(-7)(-2)=(-21)(-2)=42$

**35.** $(-3)(-2)(-3)(-4)=(6)(12)=72$

**37.** $3(-1)(5)(-6)=(-3)(-30)=90$

**39.** $(-2)(-1)(4)(-5)=2(-20)=-40$

**41.** $(-5)(4)(-3)(2)(-1)=(-20)(-6)(-1)$
$=(120)(-1)=-120$

**43.** The value of $(-2)^{13}$ is a negative number because it contains an odd number of factors.

**45.** The value of $(-96)^{52}$ is a positive number because it contains an even number of factors.

**47.** $-96^{52}$ is a negative number.

**49.** $(-10)^2=(-10)(-10)=100$

**51.** $(-5)^3=(-5)(-5)(-5)=-125$

**53.** (a) $(-4)^2=(-4)(-4)=16$
(b) $(-4)^3=(-4)(-4)(-4)=-64$

**55.** (a) $(-1)^{11}=-1$, 11 is odd
(b) $(-1)^{24}=1$, 24 is even

**57.** (a) $-4^2=-(4)(4)=-16$
(b) $(-4)^2=(-4)(-4)=16$

**59.** (a) $-2^3=-(2)(2)(2)=-8$
(b) $(-2)^3=(-2)(-2)(-2)=-8$

**61.** (a) $(-4)^3=(-4)(-4)(-4)=-64$
(b) $-4^3=-4\cdot4\cdot4=-64$

**63.** (a) $(-9)^2=(-9)(-9)=81$
(b) $-9^2=-9\cdot9=-81$

**65.** (a) $35\div7=5$ (b) $35\div(-7)=-5$
(c) $-35\div7=-5$ (d) $-35\div-7=5$

**67.** (a) $40\div8=5$ (b) $40\div(-8)=-5$
(c) $-40\div8=-5$ (d) $-40\div(-8)=5$

**69.** $20\div(-5)=-4$

**71.** $\frac{-45}{5}=-9$

**73.** $-16\div(-8)=2$

**75.** $\frac{-49}{-7}=7$

**77.** (a) $22 \div (-2) = -11$ (b) $22(-2) = -44$

**79.** (a) $-4 \div (-2) = 2$ (b) $-4(-2) = 8$

**81.** (a) $-18 \div 2 = -9$ (b) $-18(2) = -36$

**83.** (a) $14 \div -7 = -2$ (b) $-14(7) = -98$

**85.** $x^2\Big|_{x=-1} = (-1)^2 = (-1)(-1) = 1$

**87.** $\dfrac{-x}{y}\Big|_{x=36,y=6} = \dfrac{-36}{6} = -6$

**89.** $\dfrac{-m}{-n} = \dfrac{-(-10)}{-(2)} = \dfrac{10}{-2} = -5$

**91.** (a) $-y^3 = -((-2)^3) = -(-8) = 8$
(b) $-y^4 = -((-2)^4) = -(16) = -16$

**93.** $-30(3) = -90$ m or 90 m left of 0

**95.** $350(-2) = -\$700$

**97.** $\dfrac{x}{2}\Big|_{x=6} = \dfrac{6}{2} = 3 \neq -12$, no

**99.** $\dfrac{x}{-3} = 8 \Rightarrow x = -24$

**Cumulative Review**

**101.** $2^2 + 3(5) - 1 = 4 + 15 - 1 = 19 - 1 = 18$

**103.** $2^3 + (4 \div 2 + 6) = 8 + (2 + 6) = 8 + 8 = 16$

**105.** $16(5)(42) = 3360$ radios

**2.5 Exercises**

**1.** $2 + 3(-1) \neq 5(-1) = -5$ because we must multiply $3(-1)$ before we add.

**3.** Yes, $-2^2 + 8 = -4 + 8 = 4$ because there are no parentheses around $-2$ we square 2 and not $-2$ and then take the opposite.

**5.** $-2 + 3 \cdot 5 = -2 + 15 = 13$

**7.** $5 + 7(2 - 6) = 5 + 7(-4) = 5 + (-28) = -23$

**9.** $-3 + 6(8 - 5) = -3 + 6(3) = -1 + 18 = 15$

**11.** $12 - 5(2 - 6) = 12 - 5(-4) = 12 + 20 = 32$

**13.** $5(-3)(4 - 7) + 9 = -15(-3) + 9$
$= 45 + 9$
$= 54$

**15.** $-3(6 \div 3) + 7 = -3(2) + 7 = -6 + 7 = 1$

**17.** $3(-2)(9 - 5) - 10 = -6(4) - 10$
$= -24 - 10$
$= -34$

**19.** $-24 \div 12 - 8 = -2 - 8 = -10$

**21.** $(-3)^2 + 6(-9) = 9 + (-54) = -45$

**23.** $(-2)^3 - 7(8) = -8 - 56 = -64$

**25.** $(-2)^3 + 3(-8) = -8 + (-24) = -32$

**27.** $36 \div (-6) + (-6) = -6 + (-6) = -12$

**29.** $12 - 20 \div 4(-4)^2 + 9$
$= 12 - 20 \div 4(16) + 9 = 12 - 5(16) + 9$
$= 12 - 80 + 9 = -68 + 9 = -59$

**31.** $8-2(5-2^2)+6=8-2(5-4)+6$
$=8-2(1)+6$
$=8-2+6$
$=6+6=12$

**33.** $\frac{[-50\div 2+3]}{[20-9]}=\frac{-25+3}{11}=\frac{-22}{11}=-2$

**35.** $\frac{[3^2+4(-6)]}{[-3+(-2)]}=\frac{9+(-24)}{-5}=\frac{-15}{-5}=3$

**37.** $\frac{[-12-3(-2)]}{[15-17]}=\frac{-12+6}{-2}=\frac{-6}{-2}=3$

**39.** $-16\div[-4\cdot(8\div(-2))]$
$=-16\div[-4\cdot(-4)]=-16\div 16=-1$

**41.** $-60\div[5\cdot(-2\cdot(-12\div 4))]$
$=-60\div[5\cdot(-2\cdot(-3))]=-60\div[5\cdot(6)]$
$=-60\div[30]=-2$

**43.** $30,000+3(-2000)+1000$
$=30,000+(-6000)+1000$
$=24,000+1000$
$=25,000$ ft

**45.** $14(-3)+9(1)=-42+9=-33$

**47.** $7(+3)+5(-1)+4(+1)=21-5+4$
$=+20$

**49.** $1(10)+1(5)+1(3)+(1(-1)+1(-1))+2(-1)$
$=10+5+3+(-1-1)-2$
$=15+3-2-2=18-4=14$ points

**51.**
$1(20)+1(5)+2(3)+(2(-1)+2(-1))+2(-1)$
$=20+5+6+(-2-2)-2=25+6-4-2$
$=31-6=25$ points

**53.** $\frac{[(30-15\div 3)+(-5)]}{[5-10]}=\frac{(30-5)+(-5)}{-5}$
$=\frac{25+(-5)}{-5}$
$=\frac{20}{-5}=-4$

**55.** $[(3+24)\div(-3)]\cdot[2+(-3)^2]$
$=[27\div(-3)]\cdot[2+9]=-9\cdot 11=-99$

**57.** $3+x-2(-4)=7-(-13)$
$x+3-(-8)=7+13$
$x+3+8=20$
$x+11=20$
$x=9$

**Cumulative Review**

**59.** $2(x+3)=2x+2(3)=2x+6$

**61.** $3(x-2)=3x-3(2)=3x-6$

**63.** $2x+6+3x=5x+6$

**65.** $(5\cdot 32)\div 20=160\div 20=8$ bills

**2.6 Exercises**

**1.** $-2x+5x\neq -10x^2$ because we do not multiply variables and coefficients when combining like terms, we add coefficients $-2x+5x=(-2+5)x=3x$.

**3.** $-6x+(-3\boxed{x})=-9x$

**5.** $5y+\boxed{3}xy-2y+7xy=3y+10xy$

**7.** To simplify $9x+(-3xy)$ write $9x\boxed{-}3y$.

**9.** $-6(y-1)=-6\cdot\boxed{y}-(-6)\cdot\boxed{1}=-6y\boxed{+}6$

**11.** $-8x+3x=(-8+3)x=-5x$

**13.** $2x+(-3x)=(2+(-3))x=-x$

**15.** $-5x-7x=(-5-7)x=-12x$

**17.** $-7a-(-2a)=(-7-(-2))a=-5a$

**19.** $9x+(-7x)=(9+(-7))x=2x$

**21.** $-7x+(-6x)=(-7+(-6))x=-13x$

**23.** $7a+(-9b)=7a-9b$

**25.** $-4m+(-8n)=-4m-8n$

**27.** $3x+(-y)=3x-y$

**29.** $-2a-(-3b)=-2a+3b$

**31.** (a) $2-7+3=-5+3=-2$
(b) $2x-7x+3x=-5x+3x=-2x$

**33.** (a) $3-8+4=-5+4=-1$
(b) $3x-8x+4x=-5x+4x$
$=-1x$
$=-x$

**35.** (a) $2-6+1=-4+1=-3$
(b) $2x-6x+1x=-4x+x=-3x$

**37.** $-8y+5x+2y=-8y+2y+5x$
$=-6y+5x$

**39.** $6x+2y+(-8x)=6x+(-8x)+2y$
$=-2x+2y$

**41.** $9x+3y+(-5x)=9x+(-5x)+3y$
$=4x+3y$

**43.** $-8x-4x-y=(-8-4)x-y=-12x-y$

**45.** $6x+5y-10x-2y$
$=6x-10x+5y-2y$
$=-4x+3y$

**47.** $4x+2y-6x-7=4x-6x+2y-7$
$=-2x+2y-7$

**49.** $4+3ab-2-9ab=4-2+3ab-9ab$
$=2-6ab$

**51.** $5x+5xy-9x-xy=5x-9x+5xy-xy$
$=-4x+4xy$

**53.** $8a-2ab-9-7ab+3a$
$=8a+3a-9-2ab-7ab$
$=11a-9ab-9$

**55.** $3a+2x-5a+7ax-x$
$=3a-5a+2x-x+7ax$
$=-2a+x+7ax$

**57.** $6a+7b-9a+5ab-11b=-3a-4b+5ab$

**59.** $6x+8y-9x+4xy-10y=-3x-2y+4xy$

**61.** $x+3y\big|_{x=-3,y=-2}=-3+3(-2)=-3-6=-9$

**63.** $m-6n=6-6(-3)=6+18=24$

**65.** $ab-6=(-1)(5)-6=-5-6=-11$

**67.** $\frac{(a+b)}{4}=\frac{-9+5}{4}=\frac{-4}{4}=-1$

**69.** $9t^2=9(-3)^2=9(9)=81$

**71.** $8x-x^2=8(-5)-(-5)^2$
$=-40-25=-65$

**73.** $\frac{(x^2-x)}{2}=\frac{(-4)^2-(-4)}{2}$
$=\frac{16+4}{2}=\frac{20}{2}=10$

**75.** $\frac{a-b^2}{-3}=\frac{13-2^2}{-3}=\frac{13-4}{-3}=\frac{9}{-3}=-3$

**77.** $\frac{m^2+2n}{-8}=\frac{6^2+2(-2)}{-8}$
$=\frac{36-4}{-8}$
$=\frac{32}{-8}=-4$

**79.** $-3(x+2)=-3x+(-3)(2)$
$=-3x-6$

**81.** $-9(y-1)=-9y-(-9)(1)=-9y+9$

**83.** $-2(m-2)=-2m-(-2)(2)=-2m+4$

**85.** $-1(x+5)=-1x-5=-x-5$

**87.** $4(-2+y)=4(-2)+4y=-8+4y$

**89.** $2(-4+a)=2(-4)+2a=-8+2a$

**91.** $s=v-32t=-8-32(4)=-136$ ft/sec

**93.**
$v=72-32t=72-32(1)=40>0$, rising
$v=72-32t=72-32(2)=8>0$, rising
$v=72-32t=72-32(3)=-24<0$, descending

**95.** $C=\frac{5F-160}{9}=\frac{5(14)-160}{9}=-10°\text{C}$

**97.** $\frac{x^2}{4}=\frac{8^2}{4}=\frac{64}{4}=16\neq 13$, no

**99.** False, a negative number cubed results in a negative number.

## Cumulative Review

**101.** 485 = 490 to nearest ten

**103.** $P=4s=4(7)=28$ in.

**105.** $\frac{73\text{ beats}}{\text{min}}\cdot\frac{60\text{ min}}{\text{hr}}=\frac{4380\text{ beats}}{\text{hr}}$
$\frac{4380\text{ beats}}{\text{hr}}\cdot\frac{24\text{ hr}}{\text{day}}=\frac{105{,}120\text{ beats}}{\text{day}}$

## Chapter 2 Review

**1.** negative numbers – numbers less than zero

**2.** opposites – numbers the same distance from and on opposite sides of zero

**3.** integer – whole numbers and their opposites

**4.** absolute value – the absolute value of a number is the distance between the number and zero on the number line

**5.** $-3<-1$

**6.** $5 > -5$

**7.** $4 > -4$

**8.** $-7 > -11$

**9.** __+__ A profit of \$100

**10.** __−__ A drop in temperature of $16°$.

**11.** The opposite of $-16$ is __16__.

**12.** $-(-(-8)) = -(8) = -8$

**13.** $-|-11| = -11$

**14.** $|-23| = 23,\ |15| = 15,\ 23 > 15 \Rightarrow -23$
has a larger absolute value than 15.

**15.** (a) Justin made the most money in May.
(b) Justin lost the most money in March.

**16.** (a) Justin had a net gain in January, February, and May.
(b) Justin had a net loss in March and April.

**17.** $-\$500 + (-\$200) = -\$700$

**18.** $\$100 + \$200 = \$300$

**19.** (a) $-43 + (-16) = -59$
(b) $43 + 16 = 59$

**20.** (a) $-27 + (-39) = -66$
(b) $27 + 39 = 66$

**21.** $-\$25,000 + \$15,000 = -\$10,000$
The company had a net loss.

**22.** $-\$14 + \$25 = \$11$, Terry had a net profit.

**23.** (a) $-10°F + 20°F$
(b) Positive
(c) $-10°F + 20°F = 10°F$

**24.** (a) $2 + (-8) = -6$
(b) $-2 + 8 = 6$
(c) $-2 + (-8) = -10$

**25.** (a) $32 + (-18) = 14$
(b) $-32 + 18 = -14$
(c) $-32 + (-18) = -50$

**26.** $3 + (-5) + 8 + (-2) = -2 + 6 = 4$

**27.** $24 + (-52) + (-12) + (-56)$
$= -28 + (-68) = -96$

**28.** $x + 6 = -1 + 6 = 5$

**29.** $-x + y + 2 = -(-3) + (-11) + 2$
$= 3 - 11 + 2$
$= -8 + 2 = -6$

**30.** $-900 + (220) = -680$ ft

**31.** $(-240) + 350 + 400 + (-800) = -290$ ft

**32.** $-7 - 5 = -7 + (-5)$
$= -12$

**33.** $-9 - (-4) = -9 + 4 = -5$

**34.** $-5 - 3 = -5 + (-3) = -8$

**35.** $-6 - (-2) = -6 + 2 = -4$

**36.** $-3 - 8 + 6 = -11 + 6 = -5$

**37.** $6 - (-4) + (-5) = 6 + 4 + (-5) = 5$

**38.** $-4-(-2)=-4+2=-2$

**39.** $6-9-2-8=-3-2-8=-5-8=-13$

**40.** $-6-(-9)+(-1)=-6+9+(-1)$
$=3+(-1)=2$

**41.** $y-12=-2-12=-14$

**42.** $-1-x+y=-1-(-4)+(-2)=1$

**43.** $30,000-(-20,000)=\$50,000$

**44.** $10,000-(-30,000)=\$40,000$

**45.** $2300-(-1312)=3612$ ft

**46.** (a) $6(3)=18$
(b) $6(-3)=-18$
(c) $-6(3)=-18$
(d) $-6(-3)=18$

**47.** (a) $5(2)=10$
(b) $5(-2)=-10$
(c) $-5(2)=-10$
(d) $-5(-2)=10$

**48.** $-7(-5)=35$

**49.** $-2(5)=-10$

**50.** $3(-4)=-12$

**51.** $-4(-1)=4$

**52.** $(-2)(-5)(-9)=10(-9)=-90$

**53.** $(-2)(-8)(-1)(-4)=16(4)=64$

**54.** $(-5)(1)(-2)(4)(-6)=(-5)(-8)(-6)$
$=40(-6)=-240$

**55.** $(-7)^2=(-7)(-7)=49$

**56.** $-7^2=-(7)(7)=-49$

**57.** $(-6)^3=(-6)(-6)(-6)=-216$

**58.** (a) $49\div 7=7$
(b) $49\div(-7)=-7$

**59.** (a) $-30\div 5=-6$
(b) $-30\div(-5)=6$

**60.** (a) $-44\div(-4)=11$
(b) $9(-5)=-45$
(c) $(-11)(-3)=33$
(d) $\frac{25}{-5}=-5$

**61.** (a) $12\div(-4)=-3$
(b) $5(-8)=-40$
(c) $-12(-2)=24$
(d) $\frac{45}{-9}=-5$

**62.** $y^4=(-1)^4=(-1)(-1)(-1)(-1)=1$

**63.** $x^3=(-3)(-3)(-3)=-27$

**64.** $\frac{-a}{b}=\frac{-(-12)}{4}=\frac{12}{4}=3$

**65.** $\frac{-m}{-n}=\frac{-6}{-(-2)}=\frac{-6}{2}=-3$

**66.** $4-1(6-9)=4-(6-9)=4-6+9$
$=-2+9=7$

**67.** $3(-5)(2-6)+8=-15(-4)+8$
$=60+8=68$

**68.** $-2^2+3(-4)=-4+(-12)=-16$

**69.** $\frac{[-32\div 8+4]}{[7-9]}=\frac{-4+4}{-2}=\frac{0}{-2}=0$

**70.** $12+3(-5)+(-2)=12+(-15)+(-2)$
$=-3+(-2)=-5°\text{F}$

**71.** $-5y+3y=(-5+3)y=-2y$

**72.** $-4y+3x+9y=(-4+9)y+3x$
$=5y+3x=3x+5y$

**73.** $-3a-a=(-3-1)a=-4a$

**74.** $7x+9y-6x-11y=(7-6)x+(9-11)y$
$=x+(-2)y$
$=x-2y$

**75.** $3+5z-7+2yz-8z$
$=3-7+(5-8)z+2yz=-4-3z+2yz$

**76.** $-8+7y+7-2y=-8+7+(7-2)y$
$=-1+5y$

**77.** $a+3b=-1+3(-4)$
$=-1-12=-13$

**78.** $2x-y=2(-2)-(-1)$
$=-4+1=-3$

**79.** $\frac{x^2-y}{4}=\frac{(-1)^2-(-7)}{4}=\frac{8}{4}=2$

**80.** $a^2-b=(-3)^2-9=9-9=0$

**81.** $C=\frac{5F-160}{9}=\frac{5(41)-160}{9}=5°\text{F}$

**82.** $C=\frac{5F-160}{9}=\frac{5(-4)-160}{9}=-20°\text{F}$

**83.** $-6(x+2)=-6x+(-6)(2)=-6x-12$

**84.** $-2(a-1)=-2a-(-2)(1)=-2a+2$

**85.** $4(-2+x)=4(-2)+4x=-8+4x$

**How Am I Doing? Chapter 2 Test**

**1.** $-234<-5$

**2.** $|4|=4,\ |-18|=18,\ 4<18\Rightarrow |4|<|-18|$

**3.** $\underline{\ -\ }$ 14 points

**4.** $-(-(-2))=-(2)=-2$

**5.** The opposite or 10 is $\underline{-10}$.

**6.** (a) $|12|=12$  (b) $-|-3|=-(3)=-3$

**7.** (a) $-10°\text{F}+15°\text{F}$
(b) $-10°\text{F}+15°\text{F}=5°\text{F}$

**8.** $-6+8=2$

**9.** $-6+(-4)=-10$

**10.** $-20+5+(-1)+(-3)=-15+(-1)+(-3)$
$=-16+(-3)=-19$

**11.** $12-18=12+(-18)=-6$

**12.** $-1-11=-1+(-11)=-12$

**13.** $3-(-10)=3+10=13$

**14.** $-14-3+(-6)-1=-17+(-6)-1$
$=-23-1$
$=-24$

**15.** $(7)(-3)=-21$

**16.** $(-8)(-4)=32$

**17.** $(-5)(-2)(-1)(3)=10(-3)=-30$

**18.** (a) $(-5)^2=(-5)(-5)=25$
(b) $(-5)^3=(-5)(-5)(-5)=-125$
(c) $-5^2=-(5)(5)=-25$

**19.** (a) $-8\div 2=-4$
(b) $-8\div(-2)=4$

**20.** $\frac{-22}{11}=-2$

**21.** $2-35\div 5(-3)^2-6=2-7(-3)^2-6$
$=2-7(9)-6$
$=2-63-6$
$=-61-6=-67$

**22.** $\frac{[-8+2(-3)]}{[14-21]}=\frac{-8+(-6)}{-7}=\frac{-14}{-7}=2$

**23.** (a) $-7-x+y=-7-(-6)+(-3)$
$=-7+6+(-3)=-4$
(b) $-7-x+y=-7-(-7)+6$
$=0+6=6$

**24.** $\frac{2x-y^2}{-9}=\frac{2(-1)-(-4)^2}{-9}$
$=\frac{-2-16}{-9}=\frac{-18}{-9}=2$

**25.** (a) $x^4=(-1)^4=(-1)(-1)(-1)(-1)=1$
(b) $a^3=(-2)^3=(-2)(-2)(-2)=-8$

**26.** $\frac{-x}{y}=\frac{-(-6)}{-2}=\frac{6}{-2}=-3$

**27.** $5x+2y-8x-6y=(5-8)x+(2-6)y$
$=-3x-4y$

**28.** $-3x+7xy+8y-12x-11y$
$=-15x+7xy-3y$

**29.** $-6(a+7)=-6a+(-6)(7)=-6a-42$

**30.** $-2(x-1)=-2x-(-2)(1)=-2x+2$

**31.** $20,000+(-5000)=\$15,000$ profit

**32.** $3700-(-592)=4292$ ft

**33.** $s=v-32t=-7-32(5)=-167$ ft/sec

## Cumulative Test for Chapters 1 - 2

**1.** $5280=5300$ to nearest hundred

**2.** $1650-1475=175$ years

**3.** (a) $167,350-86,000=\$81,350$
(b) $81,350\div 2=\$40,675$

**4.** $6(7x)=42x$

**5.** $3(y\cdot 8)=3(8y)=24y$

**6.** $400(7n)=2800n$

**7.** $(219)(67)=14,673$

**8.** $2172 \div 14 = 155 \text{ R2}$

$$\begin{array}{r} 155 \\ 14\overline{)2172} \\ \underline{14}\phantom{00} \\ 77\phantom{0} \\ \underline{70}\phantom{0} \\ 72 \\ \underline{70} \\ 2 \end{array}$$

**9.** $3 + x = 3 + 10 = 13$

**10.** $\frac{x}{2} = 4 \Rightarrow x = 8$, check: $\frac{8}{2} \stackrel{?}{=} 4,\ 4 = 4$

**11.** Double some number equals 28:
$2x = 28$

**12.** $5 + x = 8 \Rightarrow x = 3$

**13.** $-8 > -617$

**14.** $|-17| = 17,\ |-2| = 2,\ 17 > 2 \Rightarrow |-17| > |-2|$

**15.** $-(-(-(-6))) = -(-(6))$
$= -(-6)$
$= 6$

**16.** $-|-1| = -(-(-1))$
$= -(1)$
$= -1$

**17.** $5 + (-6) = -1$

**18.** $-10 - 8 = -10 + (-8)$
$= -18$

**19.** $-7 - 6 - 4 - 8 = -13 - 4 - 8$
$= -17 - 8$
$= -25$

**20.** $(-18) \div (-9) = 2$

**21.** $(-2)^5 = (-2)(-2)(-2)(-2)(-2)$
$= -32$

**22.** $(-2)(-1)(4)(3)(-2) = 2(12)(-2)$
$= -48$

**23.** $-4 + 15 \div 5(-3)^2 - 1 = -4 + 3(9) - 1$
$= -4 + 27 - 1$
$= 23 - 1$
$= 22$

**24.** $3mn - 7mn + 4m = (3 - 7)mn + 4m$
$= -4mn + 4m$

**25.** $x - 2y^3 + 1 = -1 - 2(-3)^3 + 1$
$= -1 - 2(-27) + 1$
$= -1 - (-54) + 1$
$= -1 + 54 + 1 = 54$

# Chapter 3

## 3.1 Exercises

**1.** The sum of two opposite numbers is equal to zero.

**3.** To solve $x-6=2$, we add 6 to both sides of the equation.

**5.** $4+\boxed{-4}=0$

**7.** $-9+\boxed{9}=0$

**9.** $17+\boxed{-17}=0$

**11.** $-28+\boxed{28}=0$

**13.** $x+8+\boxed{-8}=x$

**15.** $m-2+\boxed{2}=m$

**17.**
$$\begin{array}{rrcr} & x \;+12 & = & 16 \\ + & \boxed{-12} & & \boxed{-12} \\ \hline & x \;+\boxed{0} & = & \boxed{4} \\ & x & = & \boxed{4} \end{array}$$

**19.**
$$\begin{array}{rrcr} & y \;-16 & = & 32 \\ + & \boxed{16} & & \boxed{16} \\ \hline & y \;+\boxed{0} & = & \boxed{48} \\ & y & = & \boxed{48} \end{array}$$

**21.** (a) $x-8=22,\ x-8+8=22+8,\ x=30$

check: $30-8\stackrel{?}{=}22,\ 22=22$

(b) $x+8=22,\ x+8-8=22-8,\ x=14$

check: $14+8\stackrel{?}{=}22,\ 22=22$

**23.** (a) $x+6-6=-11-6,\ x=-17$

check: $-17+6\stackrel{?}{=}-11,\ -11=-11$

(b) $x-6+6=-11+6,\ x=-5$

check: $-5-6\stackrel{?}{=}-11,\ -11=-11$

**25.** (a) $-18-2=x+2-2,\ x=-20$

check: $-18\stackrel{?}{=}-20+2,\ -18=-18$

(b) $-18+2=x-2+2,\ x=-16$

check: $-18\stackrel{?}{=}-16-2,\ -18=-18$

**27.** $y-15=5 \Rightarrow y-15+15=5+15$
$y+0=20 \Rightarrow y=20$

check: $20-15\stackrel{?}{=}5,\ 5=5$

**29.** $n-43=-74 \Rightarrow n-43+43=-74+43$
$n+0=-31 \Rightarrow n=-31$

check: $-31-43\stackrel{?}{=}-74,\ -74=-74$

**31.** $y+44=-50 \Rightarrow y+44-44=-50-44$
$y+0=-94 \Rightarrow y=-94$

check: $-94+44\stackrel{?}{=}-50,\ -50=-50$

**33.** $38+x=4 \Rightarrow -38+38+x=-38+4$
$0+x=-34 \Rightarrow x=-34$

check: $38+(-34)\stackrel{?}{=}4,\ 4=4$

**35.** $1=x-13 \Rightarrow 1+13=x-13+13$
$14=x+0 \Rightarrow x=14$

check: $1\stackrel{?}{=}14-13,\ 1=1$

**37.** $29 = y+11 \Rightarrow 29-11 = y+11-11$
$18 = y+0 \Rightarrow y = 18$

check: $29 \overset{?}{=} 18+11,\ 29 = 29$

**39.** $-13 = x+1 \Rightarrow -13-1 = x+1-1$
$-14 = x+0 \Rightarrow x = -14$

check: $-13 \overset{?}{=} -14+1,\ -13 = -13$

**41.** $4x-3x-3 = 8,\ x-3+3 = 8+3$
$x+0 = 11,\ x = 11$

check: $4(11)-3(11)-3 \overset{?}{=} 8,\ 8 = 8$

**43.** $5y-4y-1 = 5 \Rightarrow y-1 = 5$
$y-1+1 = 5+1,\ y+0 = 6,\ y = 6$

check: $5(6)-4(6)-1 \overset{?}{=} 5,\ 5 = 5$

**45.** $5 = 2y-y+1$
$5 = y+1 \Rightarrow 5-1 = y+1-1$
$4 = y+0 \Rightarrow y = 4$

check: $5 \overset{?}{=} 2(4)-4+1,\ 5 = 5$

**47.** $-23+8+x = -2+13 \Rightarrow -15+x = 11$
$-15+x+15 = 11+15 = 26$
$x+0 = 26 \Rightarrow x = 26$

check: $-23+8+26 \overset{?}{=} -2+13,\ 11 = 11$

**49.** $4-9 = a-1+14 \Rightarrow -5 = a+13$
$-5-13 = a+13-13 = a+0 \Rightarrow a = -18$

check: $4-9 \overset{?}{=} -18-1+14,\ -5 = -5$

**51.** $-45+9+m = -6+18 \Rightarrow -36+m = 12$
$-36+36+m = +12+36 \Rightarrow 0+m = 48$
$m = 48$

check: $-45+9+48 \overset{?}{=} -6+18,\ 12 = 12$

**53.** $-1+4+x = -5+9 \Rightarrow 3+x = 4$
$3-3+x = 4-3 \Rightarrow 0+x = 1 \Rightarrow x = 1$

check: $-1+4+1 \overset{?}{=} -5+9,\ 4 = 4$

**55.** $3(7-11) = y-5 \Rightarrow 3(-4) = y-5$
$-12+5 = y-5+5 \Rightarrow -7 = y+0$
$y = -7$

check: $3(7-11) \overset{?}{=} -7-5,\ -12 = -12$

**57.** $\angle a + \angle b = 180° \Rightarrow \angle a + 86° = 180°$
$\angle a = 180° - 86° \Rightarrow \angle a = 94°$

**59.** $\angle x + \angle y = 180° \Rightarrow \angle x + 112° = 180°$
$\angle x = 180° - 112° \Rightarrow \angle x = 68°$

**61.** $\angle x + \angle y = 180° \Rightarrow \angle x + 55° = 180°$
$\angle x = 180° - 55° \Rightarrow \angle x = 125°$

**63.** (a) $\angle x = \angle y + 70°$
(b) $\angle x = \angle y + 70° \Rightarrow 125° = \angle y + 70°$
$\angle y = 125° - 70° \Rightarrow \angle y = 55°$

**65.** (a) $\angle a = \angle b - 40°$
(b) $\angle a = \angle b - 40° \Rightarrow 70° = \angle b - 40°$
$\angle b = 70° + 40° \Rightarrow \angle b = 110°$

**67.** $\angle a + \angle b = 180° \Rightarrow 66° + x + 5° = 180°$
$x = 180° - 71° = 109°$
$\angle b = x + 5° = 109° + 5° = 114°$

**69.** $2^2 + (5-9) = x + 3^3$
$4 + (-4) = x + 27$
$0 - 27 = x + 27 - 27 = x + 0$
$x = -27$

**71.** $6x+3-2x = 3x-6$
$4x+3-3x-3 = 3x-6-3x-3$
$x+0 = -9 \Rightarrow x = -9$

**73.**
$$\begin{aligned}\angle a+\angle b+\angle c&=180°\\ 40°+60°+\angle c&=180°\\ \angle c&=80°\\ \angle c+\angle d+\angle e&=180°\\ 80°+\angle d+60°&=180°\\ \angle d&=40°\\ \angle a+\angle b+\angle f&=180°\\ 40°+60°+\angle f&=180°\\ \angle f&=80°\end{aligned}$$

**Cumulative Review**

**75.** Seven times $x$: $7x$

**77.** Eight times what number equals forty?
$8n = 40$

**79.** $(15\times 2)+(8\times 1.5)+(6\times -1)+(5\times 2)$
$+(5\times 2)+(2\times -1)=54$ points

**3.2 Exercises**

**1.** To solve the equation $-22x = 66$, we undo the multiplication by <u>dividing</u> both sides of the equation by <u>$-22$</u>.

**3.** $\dfrac{5x}{\boxed{5}} = x$

**5.** $\dfrac{-2x}{\boxed{-2}} = x$

**7.** $\dfrac{6\cdot x}{\boxed{6}} = x$

**9.** $\dfrac{-1\cdot x}{\boxed{-1}} = x$

**11.** $\dfrac{3x}{3}=\dfrac{39}{3}\Rightarrow x=13$

check: $3(13)\stackrel{?}{=}39,\ 39=39$

**13.** $\dfrac{11x}{11}=\dfrac{55}{11}\Rightarrow x=5$

check: $11(5)\stackrel{?}{=}55,\ 55=55$

**15.** $\dfrac{7y}{7}=\dfrac{-21}{7}\Rightarrow y=-3$

check: $7(-3)\stackrel{?}{=}-21,\ -21=-21$

**17.** $\dfrac{5m}{5}=\dfrac{-35}{5}\Rightarrow m=-7$

check: $5(-7)\stackrel{?}{=}-35,\ -35=-35$

**19.** $\dfrac{-3y}{-3}=\dfrac{15}{-3}\Rightarrow y=-5$

check: $-3(-5)\stackrel{?}{=}15,\ 15=15$

**21.** $\dfrac{-7a}{-7}=\dfrac{49}{-7}\Rightarrow a=-7$

check: $-7(-7)\stackrel{?}{=}49,\ 49=49$

**23.** $\dfrac{42}{6}=\dfrac{6x}{6}\Rightarrow x=7$

check: $42\stackrel{?}{=}6(7),\ 42=42$

**25.** $\dfrac{18}{9}=\dfrac{9x}{9}\Rightarrow x=2$

check: $18\stackrel{?}{=}9(2),\ 18=18$

**27.** $\dfrac{8x}{8}=\dfrac{104}{8}\Rightarrow x=13$

check: $8(13)\stackrel{?}{=}104,\ 104=104$

**29.** $\frac{-19x}{-19} = \frac{-76}{-19} \Rightarrow x = 4$

check: $-19(4) \stackrel{?}{=} -76,\ -76 = -76$

**31.** $2(3x) = 54 \Rightarrow 6x = 54 \Rightarrow x = \frac{54}{6} = 9$

check: $2(3(9)) \stackrel{?}{=} 54,\ 54 = 54$

**33.** $5(4x) = 40,\ 20x = 40 \Rightarrow x = \frac{40}{20} = 2$

check: $5(4(2)) \stackrel{?}{=} 40,\ 40 = 40$

**35.** $5(x \cdot 2) = \frac{40}{2} \Rightarrow 10x = 20 \Rightarrow x = \frac{20}{10} = 2$

check: $5(2 \cdot 2) \stackrel{?}{=} \frac{40}{2},\ 20 = 20$

**37.** $4(x \cdot 2) = \frac{96}{3} \Longrightarrow 8x = 32 \Rightarrow x = 4$

check: $4(4 \cdot 2) \stackrel{?}{=} \frac{96}{3},\ 32 = 32$

**39.** $-26 - 18 = 11a \Rightarrow \frac{11a}{11} = \frac{-44}{11} \Rightarrow a = -4$

check: $-26 - 18 \stackrel{?}{=} 11(-4),\ -44 = -44$

**41.** $-5 - 5 = 10y \Rightarrow \frac{-10}{10} = \frac{10y}{10} \Rightarrow y = -1$

check: $-5 - 5 \stackrel{?}{=} 10(-1),\ -10 = -10$

**43.** $8x - 2x = 24 \Rightarrow 6x = 24 \Rightarrow x = \frac{24}{6} = 4$

check: $8(4) - 2(4) \stackrel{?}{=} 24,\ 24 = 24$

**45.** $65 = 14x - 9x \Rightarrow 5x = 65 \Rightarrow x = \frac{65}{5} = 13$

check: $65 \stackrel{?}{=} 14(13) - 9(13),\ 65 = 65$

**47.** $\frac{-15y}{-15} = \frac{165}{-15} \Rightarrow y = -11$

check: $-15(-11) \stackrel{?}{=} 165,\ 165 = 165$

**49.** $2x = 3 - 11 \Rightarrow \frac{2x}{2} = \frac{-8}{2} \Rightarrow x = -4$

check: $2(-4) \stackrel{?}{=} 3 - 11,\ -8 = -8$

**51.** $11x - 3x = 56 \Rightarrow 8x = 56 \Rightarrow x = \frac{56}{8} = 7$

check: $11(7) - 3(7) \stackrel{?}{=} 56,\ 56 = 56$

**53.** $\frac{55}{5} = \frac{5a}{5} \Rightarrow a = 11$

check: $55 \stackrel{?}{=} 5(11),\ 55 = 55$

**55.** $(3x) \cdot 2 = \frac{36}{3} \Rightarrow \frac{6x}{6} = \frac{12}{6} \Rightarrow x = 2$

check: $(3 \cdot 2) \cdot 2 \stackrel{?}{=} \frac{36}{3},\ 12 = 12$

**57.** $8x + 2x = -120,\ 10x = -120,\ x = -12$

check: $8(-12) + 2(-12) \stackrel{?}{=} -120,\ -120 = -120$

**59.** $L = 4W$

**61.** $R = 2S$

**63.** (a) $R = 2B$
(b) $124{,}000 = 2B \Rightarrow B = \$62{,}000$

**65.** (a) $C = 2A$
(b) $250 = 2A \Rightarrow A = 125$ adult tickets

**67.** (a) $A = 3S$
(b) $42 = 3S \Rightarrow S = 14$ goals

**69.** (a) $W = 3M$
(b) $18 = 3M \Rightarrow M = 6$ men

**71.** $42x - 30x = 360,\ 12x = 360$
$x = 30$ shares of stock

**73.** $x + 2x = 360 \Rightarrow 3x = 360$
$x = \dfrac{360}{3} = 120$ mi

**75.** (a) $9y = 72 \Rightarrow y = \dfrac{72}{9} = 8$
(b) $-9y = 72 \Rightarrow y = \dfrac{72}{-9} = -8$
(c) $y + 9 = 72 \Rightarrow y = 72 - 9 = 63$
(d) $y - 9 = 72 \Rightarrow y = 72 + 9 = 81$

**77.** (a) $9y = -72 \Rightarrow y = \dfrac{-72}{9} = -8$
(b) $-9y = -72 \Rightarrow y = \dfrac{-72}{-9} = 8$
(c) $y + 9 = -72 \Rightarrow y = -72 - 9 = -81$
(d) $y - 9 = -72 \Rightarrow y = -72 + 9 = -63$

**79.** $x + 3x = 180° \Rightarrow 4x = 180°$
$x = \dfrac{180°}{4} = 45°,\ 3x = 135°$

**81.** $x + 9x = 180°$
$10x = 180°$
$x = 18°$
$9x = 162°$

**83.** $50° + 2x = 180° \Rightarrow 2x = 130° \Rightarrow x = 65°$

**Cumulative Review**

**85.** $P = 4s = 4(3) = 12$ in.

**87.** $LWH\big|_{L=2,W=3,H=5} = 2(3)(5) = 30$

**89.** $\$2320 - \$399 - \$118 - \$87 = \$1716$

**How Am I Doing? Sections 3.1-3.2**

**1.** $y - 15 + 15 = -26 + 15 \Rightarrow y = -11$

**2.** $4x - 3x - 2 = 9 \Rightarrow x = 9 + 2 = 11$

**3.** $2 - 8 = a - 1 + 11 \Rightarrow -6 = a + 10 \Rightarrow a = -16$

**4.** $\angle x + \angle y = 180° \Rightarrow \angle x + 115° = 180°$
$\angle x = 65°$

**5.** (a) $\angle x = \angle y + 30°$
(b) $90° = \angle y + 30°,\ \angle y = 60°$

**6.** $-16 - 12 = 14a \Rightarrow 14a = -28 \Rightarrow a = -2$

**7.** $2(3x) = -18 \Rightarrow 6x = -18 \Rightarrow x = -3$

**8.** $-48 - 10 = 8x - 6x,\ 2x = -58,\ x = -29$

**9.** $L = 4W$

**10.** (a) $C = 3A$
(b) $C = 3A \Rightarrow 150 = 3A \Rightarrow A = 50$

**11.** $65x - 50x = 795 \Rightarrow 15x = 795 \Rightarrow x = 53$
Linda purchased 53 shares.

**3.3 Exercises**

**1.** Volume, because we want to find the amount of space inside the pool.

**3.** Perimeter of a rectangle: $\underline{P = 2L + 2W}$
Area of a rectangle: $\underline{A = LW}$
Perimeter of a square: $\underline{P = 4s}$
Area of a square: $\underline{A = s^2}$
Volume of a rectangular solid: $\underline{V = LWH}$
Area of a parallelogram: $\underline{A = bh}$

**5.** (a) $P = 2L + 2W = 2(2) + 2(7)$
(b) $P = 18$ ft

**7.** (a) $P = 4s = 4(8)$
(b) $P = 32$ ft

**9.** (a) $P = 4s = 4(54)$
(b) $P = 216$ yd

**11.** (a) $P = 4s \Rightarrow 36 = 4s$
(b) $s = \frac{36}{4} = 9$ ft

**13.** (a) $P = 2L + 2W = 2(4W) + 2W = 10W$
(b) $10W = 30 \Rightarrow W = \frac{30}{10} = 3$ ft

**15.** $P = 2L + 2W = 2(10W) + 2W = 22W$
$22W = 44 \Rightarrow W = \frac{44}{22} = 2$ ft
$L = 10W = 10(2) = 20$ ft

**17.** (a) 50 squares (b) 50 $\text{in.}^2$

**19.** 40 squares

**21.** (a) $A = LW = 22(18)$
(b) $A = 396\ \text{ft}^2$

**23.** (a) $A = s^2 = 8^2$
(b) $A = 64\ \text{in.}^2$

**25.** (a) $A = bh = 12(9)$
(b) $A = 108\ \text{ft}^2$

**27.** $A = 60 = 10x \Rightarrow x = \frac{60}{10} = 6$ ft

**29.** $A = 88 = 8h \Rightarrow h = \frac{88}{8} = 11$ m

**31.** $A = 13(19) + 4(3) = 259\ \text{in.}^2$

**33.** $A = 22(9) + 8(22 - 16) + 7(14) = 344\ \text{m}^2$

**35.** (a) $4(5)(2) = 40$ cubes
(b) 40 $\text{in.}^3$

**37.** $V = LWH = 2(4)(3) = 24\ \text{ft}^3$

**39.** (a) $V = LWH = 6(5)(9)$
(b) $V = 270\ \text{in.}^3$

**41.** (a) $V = LWH = 27(10)(16)$
(b) $V = 4320\ \text{yd}^3$

**43.** $V = 300 = 10W(6) \Rightarrow W = \frac{300}{60} = 5$ cm

**45.** $P = 4s = 4(7) = 28$ in.

**47.** $A = bh = 17(8) = 136\ \text{m}^2$

**49.** $V = LWH = 12(6)(4) = 288\ \text{yd}^3$

**51.** $A = bh \Rightarrow 30 = 6x \Rightarrow x = 5$ ft

**53.** $V = 200 = L(5)(8) \Rightarrow L = \frac{200}{40} = 5$ m

**55.** (a) $2\text{ ft}\left(\frac{12\text{ in.}}{\text{ft}}\right) = 24\text{ in.}$

Dimensions are 24 in. by 24 in.

(b) $A = s^2 = 24^2 = 576\text{ in.}^2$

**57.** (a) $9\text{ ft}\left(\frac{\text{yd}}{3\text{ ft}}\right) = 3\text{ yd},\ 6\text{ ft}\left(\frac{\text{yd}}{3\text{ ft}}\right) = 2\text{ yd}$

Dimensions are 3 yd by 2 yd.

(b) $A = LW = 3(2) = 6\text{ yd}^2$

**59.** (a) $12\text{ ft}\left(\frac{\text{yd}}{3\text{ft}}\right) = 4\text{ yd},\ 9\text{ ft}\left(\frac{\text{yd}}{3\text{ft}}\right) = 3\text{ yd}$

$A = LW = 4(3) = 12\text{ yd}^2$

(b) $12\text{ yd}^2\left(\frac{\$8}{\text{yd}^2}\right) = \$96$

**61.** (a) $A = 21\text{ ft}\left(\frac{\text{yd}}{3\text{ ft}}\right)\cdot 15\text{ ft}\left(\frac{\text{yd}}{3\text{ ft}}\right) = 35\text{ yd}^2$

(b) $35\text{ yd}^2\left(\frac{\$35}{\text{yd}^2}\right) = \$560$

**63.** $(5\text{ yd})(4\text{ yd})\left(\frac{9\text{ ft}^2}{\text{yd}^2}\right)\left(\frac{\text{bag}}{100\text{ ft}^2}\right)$

$= 1.8\text{ bags} \rightarrow 2\text{ bags}\left(\frac{\$3}{\text{bag}}\right) = \$6$

**65.** $41 = x + 13 + 15 = x + 28,\ x = 13\text{ cm}$

**67.** $A = 5(7) - 3(2) = 35 - 6 = 29\text{ in.}^2$

**69.** $16(8) - 2(4) + 22(8) - 6(7) + 16(8)$

$-4(3) + 22(8) - 7(3) + 22(16) = 877\text{ ft}^2$

$877 = 2(400) + 77 \rightarrow 2\text{ gal and }1\text{ qt}$

## Cumulative Review

**71.** $3426 + 2{,}510{,}777 = 2{,}514{,}203$

$2{,}514{,}203 = 2{,}514{,}000$ to nearest thousand

**73.** $(2)(3x)(5) = 6x(5) = 5(6x) = 30x$

**75.** $-8°\text{F} + 21°\text{F} = 13°\text{F}$

## Putting Your Skills to Work, Problems for Individual Analysis

**1.** $(14\text{ ft})(11\text{ ft})\left(\frac{\$9}{\text{ft}^2}\right) = \$1386$

**2.** $(6\text{ ft})(7\text{ ft})\left(\frac{\$6}{\text{ft}^2}\right) = \$252$

## Putting Your Skills to Work, Problems for Group Investigation and Study

**1.** (a) $18(10) + 4(4) = 196\text{ ft}^2$

(b) $14(11) + 6(7) + 4(3) = 208\text{ ft}^2$

(c) $13(9) + [15(11) + 20(13)]$

$+10(11) = 652\text{ ft}^2$

(d) $652\text{ ft}^2\left(\frac{\$8}{\text{ft}^2}\right) = \$5216$

**2.** (a) $196\text{ ft}^2\left(\frac{\text{yd}^2}{9\text{ ft}^2}\right) = 21.\overline{7}\text{ yd}^2 = 20\text{ yd}^2$ to the nearest ten $\text{yd}^2$.

(b) $20\text{ yd}^2\left(\frac{\$22}{\text{yd}^2}\right) = \$440$

(c) $208\text{ ft}^2\left(\frac{\$3}{\text{ft}^2}\right) = \$624$

**3.** $\$5216 + \$440 + \$624 = \$6280$

**3.4 Understanding the Concept**

(a) $3x+5x=8x$ (b) $(3x)(5x)=15x^2$
(c) $7xy^2-5xy^2=2xy^2$
(d) $(7xy^2)(5xy^2)=35x^2y^4$

**3.4 Exercises**

**1.** No, because we do not multiply the bases.

**3.** (a) $2(3x^2)\neq 2\cdot 3+2\cdot x^2=6+2x^2$ because the distributive property is only used when the expression inside the parentheses is being added or subtracted.

(b) $2(3x^2)=2\cdot 3x^2=6x^2$ because the parentheses means multiply.
(c) $2(3+x^2)=2\cdot 3+2x^2=6+2x^2\neq 12x^2$ because 6 and $2x^2$ are not like terms and cannot be combined.

**5.** In the algebraic expression $4x^2$ the number 4 is called the coefficient or numerical coefficient.

**7.** (a) $(z\cdot z\cdot z)(z\cdot z)=z^3\cdot z^2=z^{3+2}=z^5$
(b) $z^{3+2}=z^5$

**9.** (a) $(x\cdot x)(x\cdot x\cdot x\cdot x)=x^2\cdot x^4=x^{2+4}=x^6$
(b) $x^{2+4}=x^6$

**11.** (a) $(z\cdot z)z=z^2\cdot z=z^{2+1}=z^3$
(b) $z^2\cdot z=z^{2+1}=z^3$

**13.** $2^4\cdot 2^2=2\cdot 2\cdot 2\cdot 2\cdot 2\cdot 2=2^6$

**15.** $3^5\cdot 3^3=3\cdot 3\cdot 3\cdot 3\cdot 3\cdot 3\cdot 3\cdot 3=3^8$

**17.** $x^7\cdot x^2=x^{7+2}=x^9$

**19.** $x^4\cdot x=x^{4+1}=x^5$

**21.** $3^2\cdot 3^3=3^{2+3}=3^5$

**23.** $4\cdot 4^5=4^{1+5}=4^6$

**25.** $8^2\cdot 7^5=8^2\cdot 7^5$

**27.** $x^5\cdot y^3=x^5y^3$

**29.** $x^5\cdot x^2\cdot x^7=x^{5+2+7}=x^{14}$

**31.** $y^2\cdot y^4\cdot y^5=y^{2+4+5}=y^{11}$

**33.** $3^3\cdot 3^2\cdot 3^5=3^{3+2+5}=3^{10}$

**35.** $2^5\cdot 3^2\cdot 4^7=2^5\cdot 3^2\cdot 4^7$

**37.** $(4y^5)(6y^7)=24y^{5+7}=24y^{12}$

**39.** $(6a^6)(9a^8)=54a^{6+8}=54a^{14}$

**41.** $(-3x)(8x)=-24x^2$

**43.** $(-4y)(3y)=-12y^2$

**45.** $(9a)(8a^3)(2a^6)=144a^{1+3+6}=144a^{10}$

**47.** $(x)(4x^6)(7x^5)=28x^{1+6+5}=28x^{12}$

**49.** $(5x)(3y)(-2x)=-30x^2y$

**51.** $(6y)(-3x)(-3y)=6(-3)(-3)xy^2=54xy^2$

**53.** (a) $5z^3+4$: binomial (b) 9: monomial
(c) $2x^7-3x^4-3$: trinomial

**55.** $2x(x^3+6)=2x^4+12x$

**57.** $6x^2(3x^3-1)=18x^5-6x^2$

**59.** $5y^3(y^4-2y)+3y^7=8y^7-10y^4$

**61.** $-2x^3(x^2+4x)+5x^5=3x^5-8x^4$

**63.** $(2y-5)(-6y^2)=-12y^3+30y^2$

**65.** $(2x^4-4x)(7x^2)=14x^6-28x^3$

**67.** $A=(2x^4-5)(x^3)=2x^7-5x^3$

**69.** $A=(7x^3-4)(x^2)=7x^5-4x^3$

**71.** $V=2x^4(7x^3)(x^2)=14x^9$

**73.** $P=2(4x^3)+2(3x^2)=8x^3+6x^2$

**75.** $P=2(5x^3+3x^2)+2(2x^2)$
$P=10x^3+6x^2+4x^2=10x^3+10x^2$

**77.** $A=2x^2(3x^2+5)=6x^4+10x^2$
$P=2(3x^2+5)+2(2x^2)$
$P=6x^2+10+4x^2=10x^2+10$

**79.** (a) $4cd+9cd=(4+9)cd=13cd$
(b) $(9cd)(4cd)=36c^2d^2$
(c) $4(cd+9)=4cd+36$

**81.** (a) $9ab^5-7ab^5=(9-7)ab^5=2ab^5$
(b) $(9ab^5)(7ab^5)=63ab^{10}$
(c) $9(ab^5+7)=9ab^5+63$

**83.** $W=4+L^2$

**Cumulative Review**

**85.** $700-18=682$

**87.** $20,566 \div 312 = 65$ R 286

**89.** $-1+2+0=1$ over par

**Chapter 3 Review Problems**

**1.** adjacent angles – two angles that share a common side

**2.** supplementary angles – adjacent angles formed by intersecting lines

**3.** parallel lines – straight lines that are always the same distance apart

**4.** polynomials – variable expressions that contain terms with only whole number exponents and no variables in the denominator

**5.** numerical coefficient – a number that is multiplied by a variable

**6.** monomial – a polynomial with one term

**7.** binomial – a polynomial with two terms

**8.** trinomial - a polynomial with three terms

**9.** $x-9=12 \Rightarrow x=12+9=21$

check: $21-9 \stackrel{?}{=} 12,\ 12=12$

**10.** $x+3=-7 \Rightarrow x=-7-3=-10$

check: $-10+3 \stackrel{?}{=} -7,\ -7=-7$

**11.** $-7=y+2 \Rightarrow y=-7-2=-9$

check: $-7 \stackrel{?}{=} -9+2,\ -7=-7$

**12.** $6-13+y=-4+1 \Rightarrow -7+y=-3$
$y=-3+7=4$

check: $6-13+4 \stackrel{?}{=} -4+1,\ -3=-3$

**13.** $4(2-6)=x-2 \Rightarrow 4(-4)=x-2$
$-16=x-2 \Rightarrow x=-16+2=-14$

check: $4(2-6)\overset{?}{=}-14-2,\ -16=-16$

**14.** $4x-3x-6=6 \Rightarrow x=6+6=12$

check: $4(12)-3(12)-6\overset{?}{=}6,\ 6=6$

**15.** $\angle a+\angle b=180° \Rightarrow \angle a+81°=180°$
$\angle a=180°-81°=99°$

**16.** $\angle x+\angle y=180° \Rightarrow 21°+\angle y=180°$
$\angle y=180°-21°=159°$

**17.** (a) $\angle x=\angle y+22°$
(b) $101°=\angle y+22°,\ \angle y=101°-22°=79°$

**18.** $\frac{4y}{4}=\frac{48}{4} \Rightarrow y=12$

check: $4(12)\overset{?}{=}48,\ 48=48$

**19.** $\frac{-8y}{-8}=\frac{56}{-8} \Rightarrow y=-7$

check: $-8(-7)\overset{?}{=}56,\ 56=56$

**20.** $\frac{6a}{6}=\frac{-42}{6} \Rightarrow a=-7$

check: $6(-7)\overset{?}{=}-42,\ -42=-42$

**21.** $5(3x)=45 \Rightarrow \frac{15x}{15}=\frac{45}{15} \Rightarrow x=3$

check: $5(3\cdot 3)\overset{?}{=}45,\ 45=45$

**22.** $3(y\cdot 4)=24 \Rightarrow \frac{12y}{12}=\frac{24}{12} \Rightarrow y=2$

check: $3(2\cdot 4)\overset{?}{=}24,\ 24=24$

**23.** $6(x\cdot 3)=\frac{72}{2} \Rightarrow \frac{18x}{18}=\frac{36}{18} \Rightarrow x=2$

check: $6(2\cdot 3)\overset{?}{=}\frac{72}{2},\ 36=36$

**24.** $\frac{-48}{2}=2(3\cdot x) \Rightarrow \frac{-24}{6}=\frac{6x}{6} \Rightarrow x=-4$

check: $\frac{-48}{2}\overset{?}{=}2(3(-4)),\ -24=-24$

**25.** $-4(5x)=60 \Rightarrow \frac{-20x}{-20}=\frac{60}{-20} \Rightarrow x=-3$

check: $-4(5(-3))\overset{?}{=}60,\ 60=60$

**26.** $7(2y)=28 \Rightarrow \frac{14y}{14}=\frac{28}{14} \Rightarrow y=2$

check: $7(2\cdot 2)\overset{?}{=}28,\ 28=28$

**27.** $6x-4x=20 \Rightarrow \frac{2x}{2}=\frac{20}{2} \Rightarrow x=10$

check: $6\cdot 10-4\cdot 10\overset{?}{=}20,\ 20=20$

**28.** $5x+2x=42 \Rightarrow \frac{7x}{7}=\frac{42}{7} \Rightarrow x=6$

check: $5\cdot 6+2\cdot 6\overset{?}{=}42,\ 42=42$

**29.** (a) $L=3W$ (b) $\frac{30}{3}=\frac{3W}{3} \Rightarrow W=10$ ft

**30.** (a) $W=2B$ (b) $W=2B=2(100)=200$

**31.** $(7+7+9)x=1150 \Rightarrow 23x=1150$
$x=\frac{1150}{23}=\$50$ per hour

**32.** $x+2x=12 \Rightarrow 3x=12 \Rightarrow x=\frac{12}{3}=4$ mi

**33.** $P = 28 = x + 8 + 12 \Rightarrow x + 20 = 28$
$x = 28 - 20 \Rightarrow x = 8$ in.

**34.** $P = 2L + 2W \Rightarrow 32 = 2(3W) + 2W$
$8W = 32 \Rightarrow W = \frac{32}{8} \Rightarrow W = 4$ ft

**35.** $A = s^2 = 6^2 = 36 \text{ ft}^2$

**36.** $A = LW = 72(52) = 3744 \text{ in.}^2$

**37.** $A = bh = 9(11) = 99 \text{ in.}^2$

**38.** $A = 24(11) + 10(7) + 9(16)$
$A = 478 \text{ m}^2$

**39.** $A = 104 = 13x \Rightarrow x = \frac{104}{13} = 8$ cm

**40.** $A = 96 = x \cdot 8 \Rightarrow x = \frac{96}{8} = 12$ in.

**41.** $V = 108 = 9(6) \cdot H = 54H \Rightarrow H = \frac{108}{54}$
$H = 2$ ft

**42.** $V = 198 = 11 \cdot W \cdot 6 = 66W \Rightarrow W = \frac{198}{66}$
$W = 3$ in.

**43.** $V = LWH = 25(15)(2) = 750 \text{ ft}^3$

**44.** $V = s^3 = 5^3 = 125 \text{ in.}^3$

**45.** $(16 \text{ yd})(12 \text{ yd})\left(\frac{9 \text{ ft}^2}{\text{yd}^2}\right) = 1728 \text{ ft}^2$
$1728 \text{ ft}^2\left(\frac{\text{bag}}{1000 \text{ ft}^2}\right) = 1.728 \approx 2$ bags
$\left(\frac{\$4}{\text{bag}}\right)(2 \text{ bags}) = \$8$

**46.** (a) $(18 \text{ ft})(12 \text{ ft})\left(\frac{\text{yd}^2}{9 \text{ ft}^2}\right) = 24 \text{ yd}^2$
(b) $24 \text{ yd}^2\left(\frac{\$15}{\text{yd}^2}\right) = \$360$

**47.** $a^4 \cdot a^5 = a^{4+5} = a^9$ **48.** $3^4 \cdot 3^6 = 3^{4+6} = 3^{10}$

**49.** $2^4 \cdot 4^3 = 2^4 \cdot 4^3$ **50.** $x^4 \cdot x = x^{4+1} = x^5$

**51.** $(3y^4)(5y^4) = 15y^{4+4} = 15y^8$

**52.** $(4x^2)(3x^6) = 12x^{2+6} = 12x^8$

**53.** $(3a)(a^3)(7a^6) = 21a^{1+3+6} = 21a^{10}$

**54.** $(4z)(y^8)(3z^3)(2y^3) = 24zy^{8+3} = 24zy^{11}$

**55.** $(4x^5)(-3x^2) = -12x^{5+2} = -12x^7$

**56.** $(-7z^7)(5z^3) = -35z^{7+3} = -35z^{10}$

**57.** $(-3a^4)(-4a^{10}) = 12a^{4+10} = 12a^{14}$

**58.** $(3y^4)(-6y^{11}) = -18y^{4+11} = -18y^{15}$

**59.** $3x^2 - 1$: binomial **60.** $4x^3$: monomial

**61.** $2xy^2 + 3x + 1$: trinomial

**62.** $x(x^2 + 2) = x \cdot x^2 + x \cdot 2 = x^3 + 2x$

**63.** $5x(x^3 - 4) = 5x \cdot x^3 - 5x(4) = 5x^4 - 20x$

**64.** $A = LW = (3x^2 + 6) \cdot x^3 = 3x^5 + 6x^3$

**65.** $V = LWH = 2x^4 \cdot 3x^2 \cdot 2x = 12x^7$

**How Am I Doing? Chapter 3 Test**

**1.** $x-2=-8 \Rightarrow x=-8+2=-6$

check: $-6-2\stackrel{?}{=}-8,\ -8=-8$

**2.** $y+3=72 \Rightarrow y=72-3=69$

check: $69+3\stackrel{?}{=}72,\ 72=72$

**3.** $9-15=a-3 \Rightarrow a=-6+3=-3$

check: $9-15\stackrel{?}{=}-3-3,\ -6=-6$

**4.** $3x-2x-7=1+6,\ x=5+7=12$

check: $3(12)-2(12)-7\stackrel{?}{=}-1+6,\ 5=5$

**5.** $12=5x-2x \Rightarrow 3x=12 \Rightarrow x=4$

check: $12\stackrel{?}{=}5(4)-2(4),\ 12=12$

**6.** $-3y=42 \Rightarrow y=\dfrac{42}{-3}=-14$

check: $-3(-14)\stackrel{?}{=}42,\ 42=42$

**7.** $2(4x)=-72 \Rightarrow 8x=-72$

$x=\dfrac{-72}{8}=-9$

check: $2(4(-9))\stackrel{?}{=}-72,\ -72=-72$

**8.** $\dfrac{-16}{2}=2+y,\ -8=2+y$

$y=-8-2=-10$

check: $\dfrac{-16}{2}\stackrel{?}{=}2+(-10),\ -8=-8$

**9.** $\angle x+\angle y=180° \Rightarrow 75°+\angle y=180°$

$\angle y=180°-75°=105°$

**10.** (a) $\angle x=\angle y+10°$

(b) $95°=\angle y+10° \Rightarrow \angle y=95°-10°=85°$

**11.** (a) $F=3M$

(b) $21=3M \Rightarrow M=\dfrac{21}{3}=7$ male students

**12.** $P=48=2L+2W=2(3W)+2W$

$6W+2W=8W=48 \Rightarrow W=\dfrac{48}{8}=6$ yd

**13.** $A=LW=5(3)=15\text{ ft}^2$

**14.** $A=bh=6(12)=72\text{ in.}^2$

**15.** $A=15(12)-8(12-5)=124\text{ cm}^2$

**16.** $V=288=LWH=12W(8)=96W$

$W=\dfrac{288}{96}=3$ in.

**17.** $A=LW=(4x^3+1)(x^2)=4x^5+x^2$

**18.** $A=LW$

$A=(9\text{ ft})\left(\dfrac{\text{yd}}{3\text{ ft}}\right)\cdot(6\text{ ft})\left(\dfrac{\text{yd}}{3\text{ ft}}\right)=6\text{ yd}^2$

**19.** $V=LWH=30(40)(10)=12{,}000\text{ ft}^3$

**20.** $y^3\cdot y^2=y^{3+2}=y^5$

**21.** $z\cdot z^3=z^{1+3}=z^4$

**22.** $2^3\cdot 3^2=8\cdot 9=72$

**23.** $(5x)(x^3)(x^4)=5x^{1+3+4}=5x^8$

**24.** $(-8x^2)(-9x^4)=72x^{2+4}=72x^6$

**25.** $5y(y^4+8)=5y\cdot y^4+5y\cdot 8=5y^5+40y$

**26.** $6x^3(3x^2-5x)=6x^3\cdot 3x^2-6x^3\cdot 5x$
$=18x^5-30x^4$

## Cumulative Test for Chapters 1-3

**1.** $3982 = 4000$ to nearest thousand

**2.** $1650-1475=175$ years

**3.** $x+9=16$

**4.** $5+9y+3+y=5+3+9y+y=8+10y$

**5.** $(219)(67)=14,673$

**6.** $8446\div 41=206$

**7.** (a) $167,350-86,000=\$81,350$
(b) $81,350\div 2=\$40,675$

**8.** $-1>-100$

**9.** $|-3|=3,\ |-2|=2,\ 3>2\Rightarrow |-3|>|-2|$

**10.** $(5)+(-5)=0$

**11.** $-10-81=-10+(-81)=-91$

**12.** $(16)(-4)=-64$

**13.** $(-36)\div(-9)=4$

**14.** $(-1)^5=(-1)(-1)(-1)(-1)(-1)=-1$

**15.** $25+x=85\Rightarrow x=85-25=60$

**16.** $9y=63\Rightarrow y=\dfrac{63}{9}=7$

**17.** $-36+y=-2\Rightarrow y=-2+36=34$

**18.** $2x-x+5=-6\Rightarrow x=-6-5=-11$

**19.** $\dfrac{120}{10}=-3y\Rightarrow -3y=12\Rightarrow y=\dfrac{12}{-3}=-4$

**20.** $2+6=-3+4+x\Rightarrow 8=1+x$
$x=8-1=7$

**21.** $P=2L+2W=2(3)+2(2)=10$ in.

**22.** $V=LWH,\ 30=L(2)(3)\Rightarrow 6L=30$
$L=\dfrac{30}{6}=5$ m

**23.** $\angle a+\angle b=180°\Rightarrow \angle a+105°=180°$
$\angle a=180°-105°=75°$

**24.** (a) $C=2A$
(b) $140=2A\Rightarrow A=70$ adults

**25.** $(-3x^2)(4x^6)=-12x^{2+6}=-12x^8$

**26.** $(5y^2)(3y^5)=15y^{2+5}=15y^7$

**27.** $x^3(2x^2+5)=x^3\cdot 2x^2+x^3\cdot 5$
$=2x^5+5x^3$

**28.** $A=LW=(3x^3-1)(2x^2)$
$A=3x^3\cdot 2x^2-1\cdot 2x^2$
$A=6x^5-2x^2$

# Chapter 4

## 4.1 Exercises

**1.** A number is divisible by 2 if it is even.

**3.** A number is divisible by 5 if the last digit is 5 or 0.

**5.** With this method the divisors and quotient are the prime factors. Therefore, the divisor (number you divide by) must be prime for your factors to be prime.

**7.** 155 is not divisible by 2 because it is not even.

**9.** 232 is not divisible by three because the sum of the digits, $2+3+2=7$, is not divisible by 3.

**11.** 324 is even and therefore divisible by 2. The sum of the digits, $3+2+4=9$, is divisible by 3 and therefore 324 is divisible by 3. The last digit is not a 0 or 5, therefore 324 is not divisible by 5.

**13.** 805 is not even and therefore not divisible by 2. The sum of the digits, $8+0+5=13$, is not divisible by 3 and therefore 805 is not divisible by 3. The last digit is 5, therefore 805 is divisible by 5.

**15.** 330 is even and therefore divisible by 2. The sum of the digits, $3+3+0=6$, is divisible by 3 and therefore 330 is divisible by 3. The last digit is 0, therefore 330 is divisible by 5.

**17.** 22,971 is not even and therefore is not divisible by 2. The sum of the digits, $2+2+9+7+1=21$, is divisible by 3 and therefore 22,971 is divisible by 3.

**17.** The last digit is not a 0 or 5, therefore 22,971 is not divisible by 5.

**19.** 0,1: neither. 23: prime. 9,40,8,15,33: composite.

**21.** (a) $8 = 2\cdot 4 = 8\cdot 1$
(b) $8 = 2\cdot 2\cdot 2 = 2^3$

**23.** $28 = 2\cdot\boxed{2}\cdot 7 = 2^{\boxed{2}}\cdot 7$

**25.** $50 = 2\cdot\boxed{5}\cdot 5 = 2\cdot 5^{\boxed{2}}$

**27.** (a)

$$\begin{array}{r} \boxed{5} \\ 2\,\overline{)10} \\ \boxed{3}\,\overline{)30} \\ 5\,\overline{)150} \end{array}$$

(b) $150 = 2\cdot 3\cdot 5^2$

**29.** (a)

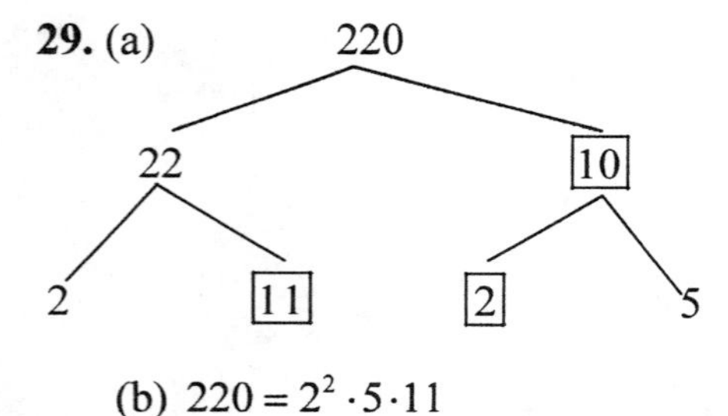


(b) $220 = 2^2\cdot 5\cdot 11$

**31.** $9 = 3\cdot 3 = 3^2$

**33.** $12 = 2\cdot 2\cdot 3 = 2^2\cdot 3$

**35.** $24 = 2\cdot 2\cdot 2\cdot 3 = 2^3\cdot 3$

**37.** $70 = 2\cdot 5\cdot 7$

**39.** $64 = 2\cdot 2\cdot 2\cdot 2\cdot 2\cdot 2 = 2^6$

**41.** $80 = 2 \cdot 2 \cdot 2 \cdot 2 \cdot 5 = 2^4 \cdot 5$

**43.** $75 = 3 \cdot 5 \cdot 5 = 3 \cdot 5^2$

**45.** $45 = 3 \cdot 3 \cdot 5 = 3^2 \cdot 5$

**47.** $99 = 3 \cdot 3 \cdot 11 = 3^2 \cdot 11$

**49.** $300 = 2^2 \cdot 3 \cdot 5^2$

**51.** $110 = 2 \cdot 5 \cdot 11$

**53.** $136 = 2 \cdot 2 \cdot 2 \cdot 17 = 2^3 \cdot 17$

**55.** $2400 = 2^5 \cdot 3 \cdot 5^2$

**57.** $3528 = 2^3 \cdot 3^2 \cdot 7^2$

**59.**

**61.**

```
prgmPRIMEFAC
PRIME FACTORS OF
?91
                7
               13
DONE
             Done
```

```
PRIME FACTORS OF
?561
                3
               11
               17
DONE
             Done
```

**63.**

```
7*13
               91
3*11*17
              561
```

**65.** For a six-digit number to be divisible by 2 and 3 it must be an even and the sum of its digits must be divisible by 3.

## Cumulative Review

**67.** $6y^2 + 3y^2 = 9y^2$

**69.** $(5x)(3x)(2) = 30x^2$

**71.** $\angle x + \angle y = 180° \Rightarrow \angle x + 62° = 180°$
$\angle x = 180° - 62° = 118°$

## 4.2 Exercises

**1.** Fractions are a set of numbers used to describe <u>part</u> of a whole quantity. In the fraction $\frac{2}{3}$ the 2 is called the <u>numerator</u> and the 3 is called the <u>deonomiator</u>.

**3.** To change the mixed number $6\frac{2}{3}$ to an improper fraction, multiply $\underline{3} \times \underline{6}$, and then add $\underline{2}$ to the product. The improper fraction is written as $\frac{\boxed{20}}{\boxed{3}}$.

**5.** $\frac{3}{5}$

**7.** $\frac{3}{4}$

**9.** $\frac{7}{0}$ undefined

**11.** $\frac{0}{z} = 0$

**13.** $\frac{44}{44} = 1$

**15.** $\frac{y}{y} = 1$

**17.** $\frac{8}{0}$ undefined

**19.** $\frac{0}{8} = 0$

**21.** $\frac{7}{15}$

**23.** $\frac{37}{37+57}=\frac{37}{94}$

**25.** $\frac{26-9}{26}=\frac{17}{26}$

**27.** $\frac{32-21}{32}=\frac{11}{32}$

**29.** $\frac{24+58}{24+22+58+31}=\frac{82}{135}$

**31.** $\frac{24+22+58}{24+22+58+31}=\frac{104}{135}$

**33.** $\frac{7}{9}$: proper fraction

**35.** $\frac{9}{9}$: improper fraction

**37.** $\frac{6}{6}$: improper fraction

**39.** $7\frac{1}{9}$: mixed number

**39.** $\frac{13}{7}=1\frac{6}{7}$

**41.** $\frac{73}{4}=18\frac{1}{4}$

**43.** $\frac{41}{2}=20\frac{1}{2}$

**45.** $\frac{32}{5}=6\frac{2}{5}$

**47.** $\frac{47}{5}=9\frac{2}{5}$

**49.** $\frac{33}{33}=1$

**51.** $8\frac{3}{7}=\frac{8(7)+3}{7}=\frac{56+3}{7}=\frac{59}{7}$

**53.** $24\frac{1}{4}=\frac{4(24)+1}{4}=\frac{96+1}{4}=\frac{97}{4}$

**55.** $15\frac{2}{3}=\frac{3(15)+2}{3}=\frac{45+2}{3}=\frac{47}{3}$

**57.** $33\frac{1}{3}=\frac{3(33)+1}{3}=\frac{99+1}{3}=\frac{100}{3}$

**59.** $8\frac{9}{10}=\frac{10(8)+9}{10}=\frac{80+9}{10}=\frac{89}{10}$

**61.** $8\frac{21}{30}=\frac{30(8)+21}{30}=\frac{240+21}{30}=\frac{261}{30}$

**Cumulative Review**

**63.** $(-5)(-8)=40$

**65.** $63=3\cdot 3\cdot 7=3^2\cdot 7$

**67.** $-71°\text{C}-(-281°\text{C})=210°\text{C}$

### 4.3 Exercises

**1.** To build an equivalent fraction we multiply the numerator and denominator by the same number.

**3.** $\frac{3}{7}\cdot\frac{\boxed{2}}{2}=\frac{\boxed{6}}{14}$

**5.** $\frac{4}{5x}\cdot\frac{\boxed{3}}{3}=\frac{\boxed{12}}{15x}$

**7.** $\frac{7}{8}\cdot\frac{\boxed{y}}{\boxed{y}}=\frac{7y}{8y}$

**9.** $\frac{2}{9}\cdot\frac{\boxed{y}}{\boxed{y}}=\frac{\boxed{2y}}{9y}$

**11.** (a) $\frac{7}{9}\cdot\frac{4}{4}=\frac{28}{36}$

(b) $\frac{7}{9}\cdot\frac{5x}{5x}=\frac{35x}{45x}$

**13.** (a) $\frac{4}{11}\cdot\frac{4}{4}=\frac{16}{44}$

(b) $\frac{4}{11}\cdot\frac{5x}{5x}=\frac{20x}{55x}$

**15.** $\frac{3}{7}\cdot\frac{7}{7}=\frac{21}{49}$

**17.** $\frac{3}{4}\cdot\frac{5}{5}=\frac{15}{20}$

**19.** $\frac{9}{13}\cdot\frac{3}{3}=\frac{27}{39}$

**21.** $\frac{35}{40}\cdot\frac{2}{2}=\frac{70}{80}$

**23.** $\frac{8}{9}\cdot\frac{y}{y}=\frac{8y}{9y}$

**25.** $\frac{3}{7}\cdot\frac{4y}{4y}=\frac{12y}{28y}$

**27.** $\frac{3}{6}\cdot\frac{3a}{3a}=\frac{9a}{18a}$

**29.** $\frac{5}{7}\cdot\frac{3x}{3x}=\frac{15x}{21x}$

**31.** $\frac{15}{25}=\frac{3\cdot\cancel{5}}{5\cdot\cancel{5}}=\frac{3}{5}$

**33.** $\frac{12}{16}=\frac{\cancel{2}\cdot\cancel{2}\cdot3}{\cancel{2}\cdot\cancel{2}\cdot2\cdot2}=\frac{3}{4}$

**35.** $\frac{30}{36}=\frac{\cancel{2}\cdot\cancel{3}\cdot5}{\cancel{2}\cdot2\cdot\cancel{3}\cdot3}=\frac{5}{6}$

**37.** $\frac{24}{28}=\frac{\cancel{2}\cdot\cancel{2}\cdot2\cdot3}{\cancel{2}\cdot\cancel{2}\cdot7}=\frac{6}{7}$

**39.** $\frac{24}{36}=\frac{\cancel{2}\cdot\cancel{2}\cdot2\cdot\cancel{3}}{\cancel{2}\cdot\cancel{2}\cdot\cancel{3}\cdot3}=\frac{2}{3}$

**41.** $\frac{30}{85}=\frac{2\cdot3\cdot\cancel{5}}{\cancel{5}\cdot17}=\frac{6}{17}$

**43.** $\frac{42}{54}=\frac{\cancel{2}\cdot\cancel{3}\cdot7}{\cancel{2}\cdot\cancel{3}\cdot3\cdot3}=\frac{7}{9}$

**45.** $\frac{36}{72}=\frac{\cancel{2}\cdot\cancel{2}\cdot\cancel{3}\cdot\cancel{3}}{\cancel{2}\cdot\cancel{2}\cdot2\cdot\cancel{3}\cdot\cancel{3}}=\frac{1}{2}$

**47.** $\frac{49}{35}=\frac{7\cdot\cancel{7}}{5\cdot\cancel{7}}=\frac{7}{5}=1\frac{2}{5}$

**49.** $\frac{75}{60}=\frac{\cancel{3}\cdot\cancel{5}\cdot5}{2\cdot2\cdot\cancel{3}\cdot\cancel{5}}=\frac{5}{4}=1\frac{1}{4}$

**51.** (a) $\frac{-12}{18} = -\frac{\cancel{2}\cdot 2\cdot \cancel{3}}{\cancel{2}\cdot \cancel{3}\cdot 3} = -\frac{2}{3}$

(b) $\frac{12}{-18} = -\frac{\cancel{2}\cdot 2\cdot \cancel{3}}{\cancel{2}\cdot \cancel{3}\cdot 3} = -\frac{2}{3}$

(c) $-\frac{12}{18} = -\frac{\cancel{2}\cdot 2\cdot \cancel{3}}{\cancel{2}\cdot \cancel{3}\cdot 3} = -\frac{2}{3}$

**53.** $\frac{-24}{36} = \frac{-\cancel{2}\cdot \cancel{2}\cdot 2\cdot \cancel{3}}{\cancel{2}\cdot \cancel{2}\cdot \cancel{3}\cdot 3} = -\frac{2}{3}$

**55.** $\frac{-42}{48} = \frac{-\cancel{2}\cdot \cancel{3}\cdot 7}{\cancel{2}\cdot 2\cdot 2\cdot 2\cdot \cancel{3}} = -\frac{7}{8}$

**57.** $\frac{30}{-42} = -\frac{\cancel{2}\cdot \cancel{3}\cdot 5}{\cancel{2}\cdot \cancel{3}\cdot 7} = -\frac{5}{7}$

**59.** $-\frac{16}{18} = -\frac{\cancel{2}\cdot 2\cdot 2\cdot 2}{\cancel{2}\cdot 3\cdot 3} = -\frac{8}{9}$

**61.** $\frac{16x}{18x} = \frac{\cancel{2}\cdot 2\cdot 2\cdot 2\cancel{x}}{\cancel{2}\cdot 3\cdot 3\cancel{x}} = \frac{8}{9}$

**63.** $\frac{20y}{35y} = \frac{2\cdot 2\cdot \cancel{5}\cancel{y}}{\cancel{5}\cdot 7\cancel{y}} = \frac{4}{7}$

**65.** $\frac{24xy}{42x} = \frac{\cancel{2}\cdot 2\cdot 2\cdot \cancel{3}\cancel{x}y}{\cancel{2}\cdot \cancel{3}\cdot 7\cancel{x}} = \frac{4y}{7}$

**67.** $\frac{21x}{28xy} = \frac{3\cdot \cancel{7}\cancel{x}}{2\cdot 2\cdot \cancel{7}\cancel{x}y} = \frac{3}{4y}$

**69.** $\frac{27x^2}{45x} = \frac{\cancel{3}\cdot \cancel{3}\cdot 3\cdot \cancel{x}\cdot x}{\cancel{3}\cdot \cancel{3}\cdot 5\cdot \cancel{x}} = \frac{3x}{5}$

**71.** $\frac{20y}{24y^2} = \frac{\cancel{2}\cdot \cancel{2}\cdot 5\cancel{y}}{\cancel{2}\cdot \cancel{2}\cdot 2\cdot 3\cdot \cancel{y}\cdot y} = \frac{5}{6y}$

**73.** $\frac{36n^2}{-42n} = -\frac{\cancel{2}\cdot 2\cdot \cancel{3}\cdot 3\cdot \cancel{n}\cdot n}{\cancel{2}\cdot \cancel{3}\cdot 7\cancel{n}} = -\frac{6n}{7}$

**75.** $\frac{-35x}{45x^2} = -\frac{\cancel{5}\cdot 7\cancel{x}}{3\cdot 3\cdot \cancel{5}\cancel{x}\cdot x} = -\frac{7}{9x}$

**77.** $\frac{36-22}{36} = \frac{14}{36} = \frac{\cancel{2}\cdot 7}{\cancel{2}\cdot 18} = \frac{7}{18}$

**79.** $\frac{22+15+20}{36+20+25} = \frac{57}{81} = \frac{19}{27}$

**81.** $\frac{60}{53+54+57+83+69+60} = \frac{60}{376} = \frac{15}{94}$

**83.** $\frac{57+83}{376} = \frac{140}{376} = \frac{35}{94}$

**85.** $\frac{40a^2b^2c^4}{88ab} = \frac{5\cdot \cancel{8}\cancel{a}\cdot a\cdot \cancel{b}\cdot bc^4}{\cancel{8}\cdot 11\cancel{a}\cancel{b}} = \frac{5abc^4}{11}$

**87.** $\frac{256xy^3}{300yz^5} = \frac{\cancel{4}\cdot 64x\cancel{y}\cdot y^2}{\cancel{4}\cdot 75\cancel{y}z^5} = \frac{64xy^2}{75z^5}$

**89.** $A = LW = 90(80) = 120W \Rightarrow W = 60$ in.

## Cumulative Review

**91.** $9-(6-1) = 9-5 = 4$

**93.** (a) $P = 2C$
(b) $34 = 2C \Rightarrow C = 17$

## How Am I Doing? Sections 4.1-4.3

**1.** $312 = 2^3\cdot 3\cdot 13$ is divisible by 2 and 3 but not 5.

**2.** $120 = 2^3\cdot 3\cdot 5$

**3.** (a) $\frac{0}{3}=0$ (b) $\frac{3}{0}$ undefined

(c) $\frac{a}{a}=1$ (d) $\frac{22}{22}=1$

**4.** $\frac{51}{7}=7\frac{2}{7}$

**5.** $5\frac{3}{4}=\frac{4\cdot 5+3}{4}=\frac{23}{4}$

**6.** $\frac{6}{30}=\frac{1}{5}$

**7.** $\frac{3}{7}\cdot\frac{5}{5}=\frac{15}{35}$

**8.** $\frac{2}{9}\cdot\frac{3y}{3y}=\frac{6y}{27y}$

**9.** $\frac{20}{-42}=-\frac{2\cdot 10}{2\cdot 21}=-\frac{10}{21}$

**10.** $\frac{25y}{45y^2}=\frac{5\cdot 5y}{9\cdot 5y\cdot y}=\frac{5}{9y}$

**11.** $\frac{42-7}{42}=\frac{35}{42}=\frac{5\cdot 7}{6\cdot 7}=\frac{5}{6}$

## 4.4 Exercises

**1.** (a) We add the coefficients and the variables stay the same.
(b) We multiply coefficients and then add exponents of like bases.
(c) We simplify coefficients and then subtract exponents of like bases.

**3.** $\frac{7^4}{7^3}=7^{4-3}=7^1=7$

**5.** $\frac{a^8}{a^3}=a^{8-3}=a^5$

**7.** $\frac{3^5}{3^9}=3^{5-9}=3^{-4}=\frac{1}{3^4}$

**9.** $\frac{3^2}{3^6}=3^{2-6}=3^{-4}=\frac{1}{3^4}$

**11.** $\frac{z^4}{y^8}=\frac{z^4}{y^8}$

**13.** $\frac{9^3}{8^8}=\frac{9^3}{8^8}$

**15.** $\frac{z^8}{z^8}=1$

**17.** $\frac{y^3z^4}{y^5z^7}=y^{3-5}z^{4-7}=y^{-2}z^{-3}=\frac{1}{y^2z^3}$

**19.** $\frac{m^9 3^6}{m^7 3^7}=m^{9-7}\cdot 3^{6-7}=m^2\cdot 3^{-1}=\frac{m^2}{3}$

**21.** $\frac{a^5 7^4}{a^3 7^7}=a^{5-3}\cdot 7^{4-7}=a^2\cdot 7^{-3}=\frac{a^2}{7^3}$

**23.** $\frac{b^9 9^9}{b^7 9^{11} 3^0}=b^{9-7}9^{9-11}=b^2 9^{-2}=\frac{b^2}{9^2}$

**25.** $\frac{4^6 ab^0}{4^9 a^4 b}=\frac{1}{4^{9-6}a^{4-1}b}=\frac{1}{4^3a^3b}$

**27.** $\frac{25y^3}{35y}=\frac{\cancel{5}\cdot 5\cancel{y}y^2}{\cancel{5}\cdot 7\cancel{y}}=\frac{5y^2}{7}$

**29.** $\frac{9a^4}{27a^3} = \frac{\cancel{9}\cancel{a^3}a}{\cancel{9}\cdot 3\cancel{a^3}} = \frac{a}{3}$

**31.** $\frac{56x^9y^0}{64x^3} = \frac{7\cdot\cancel{8}x^{9-3}\cdot 1}{8\cdot\cancel{8}} = \frac{7x^6}{8}$

**33.** $\frac{12x^4y^2}{15xy^3} = \frac{3\cdot 4x^{4-1}}{3\cdot 5y^{3-2}} = \frac{4x^3}{5y}$

**35.** (a) $z^2\cdot z^2\cdot z^2 = z^{2+2+2} = z^6$
(b) $(z^2)^3 = z^{2\cdot 3} = z^6$

**37.** (a) $x^2\cdot x^2 = x^{2+2} = x^4$
(b) $(x^2)^2 = x^{2\cdot 2} = x^4$

**39.** $(z^6)^4 = z^{6\cdot 4} = z^{24}$

**41.** $(3^3)^2 = 3^{3\cdot 2} = 3^6$

**43.** $(b^1)^6 = (b)^6 = b^6$

**45.** $(x^0)^4 = (1)^4 = 1\cdot 1\cdot 1\cdot 1 = 1$

**47.** $(y^2)^3 = y^{2\cdot 3} = y^6$

**49.** $(2^4)^5 = 2^{4\cdot 5} = 2^{20}$

**51.** $(x^2)^0 = x^{2\cdot 0} = x^0 = 1$

**53.** $(7^3)^9 = 7^{3\cdot 9} = 7^{27}$

**55.** $(x^2)^2 = x^{2\cdot 2} = x^4$

**57.** $(y^2)^6 = y^{2\cdot 6} = y^{12}$

**59.** $(4x^2)^3 = 4^3x^{2\cdot 3} = 4^3x^6$

**61.** $(3a^4)^8 = 3^8a^{4\cdot 8} = 3^8a^{32}$

**63.** $(2^2x^5)^3 = 2^{2\cdot 3}x^{5\cdot 3} = 2^6x^{15}$

**65.** $(5^3a^4)^5 = 5^{3\cdot 5}a^{4\cdot 5} = 5^{15}a^{20}$

**67.** $\left(\frac{2}{x}\right)^2 = \left(\frac{2}{x}\right)\left(\frac{2}{x}\right) = \frac{2^2}{x^2} = \frac{4}{x^2}$

**69.** $\left(\frac{y}{x}\right)^9 = \frac{y^9}{x^9}$

**71.** $\left(\frac{3}{x}\right)^3 = \frac{3^3}{x^3} = \frac{27}{x^3}$

**73.** $\left(\frac{a}{b}\right)^4 = \frac{a^4}{b^4}$

**75.** $\left(\frac{x}{6}\right)^2 = \frac{x^2}{6^2} = \frac{x^2}{36}$

**77.** $\left(\frac{3}{4}\right)^2 = \frac{3^2}{4^2} = \frac{9}{16}$

**79.** $\frac{25x^2y^3z^4}{135x^7y} = \frac{\cancel{5}\cdot 5\cancel{x^2}\cancel{y}y^2z^4}{\cancel{5}\cdot 27\cancel{x^2}x^5\cancel{y}} = \frac{5y^2z^4}{27x^5}$

**81.** $\frac{156a^0b^8}{144b^6c^9} = \frac{\cancel{12}\cdot 13\cdot 1\cdot b^{8-6}}{\cancel{12}\cdot 12c^9} = \frac{13b^2}{12c^9}$

**83.** $\left(\frac{2y}{5x}\right)^2 = \frac{2^2y^2}{5^2x^2} = \frac{4y^2}{25x^2}$

**85.** $\left(\frac{3a^2}{2b^3}\right)^3 = \frac{3^3a^{2\cdot 3}}{2^3b^{3\cdot 3}} = \frac{27a^6}{8b^9}$

**87.** (a) $15x^3 + 5x^3 = (15+5)x^3 = 20x^3$

(b) $(15x^3)(5x^3) = 75x^{3+3} = 75x^6$

(c) $(x^3)^3 = x^{3\cdot 3} = x^9$

(d) $\dfrac{15x^3}{5x^5} = 3x^{3-5} = 3x^{-2} = \dfrac{3}{x^2}$

**89.** (a) $3x^3 + 9x^3 = (3+9)x^3 = 12x^3$

(b) $(3x^3)(9x^3) = 27x^{3+3} = 27x^6$

(c) $(3x^3)^3 = 3^3x^{3\cdot 3} = 3^3x^9$

(d) $\dfrac{3x^3}{9x^4} = \dfrac{\cancel{3}\cancel{x^3}}{\cancel{3}\cdot 3\cancel{x^3}\cdot x} = \dfrac{1}{3x}$

**91.** (a) $12x^4 + 3x^4 = 15x^4$

(b) $(12x^4)(3x^4) = 12\cdot 3x^{4+4} = 36x^8$

(c) $(2x^4)^4 = 2^4x^{4\cdot 4} = 2^4x^{16} = 16x^{16}$

(d) $\dfrac{2x^4}{3x} = \dfrac{2x^{4-1}}{3} = \dfrac{2x^3}{3}$

**93.** (a) $5y^2 + 15y^2 = 20y^2$

(b) $(5y^2)(15y^2) = 5\cdot 15y^{2+2} = 75y^4$

(c) $(5y^2)^2 = 5^2y^{2\cdot 2} = 5^2y^4 = 25y^4$

(d) $\dfrac{5y^2}{15y^7} = \dfrac{5}{5\cdot 3y^{7-2}} = \dfrac{1}{3y^5}$

### Cumulative Review

**95.** $3x = 42 \Rightarrow x = \dfrac{42}{3} = 14$

**97.** $(18-4) = 7x \Rightarrow 7x = 14 \Rightarrow x = \dfrac{14}{7} = 2$

**99.** $A = LW = 2x^2(3x+3) = 6x^3 + 6x^2$

### 4.5 Exercises

**1.** A ratio compares amounts with the same units and a rate compares amounts with different units.

**3.** $\dfrac{15}{65} = \dfrac{3\cdot\cancel{5}}{\cancel{5}\cdot 13} = \dfrac{3}{13}$

**5.** $\dfrac{35}{10} = \dfrac{\cancel{5}\cdot 7}{\cancel{5}\cdot 2} = \dfrac{7}{2}$

**7.** $\dfrac{25}{70} = \dfrac{\cancel{5}\cdot 5}{\cancel{5}\cdot 14} = \dfrac{5}{14}$

**9.** $\dfrac{34 \text{ min}}{12 \text{ min}} = \dfrac{\cancel{2}\cdot 17}{\cancel{2}\cdot 6} = \dfrac{17}{6}$

**11.** $\dfrac{14 \text{ gal}}{35 \text{ gal}} = \dfrac{2\cdot\cancel{7}}{5\cdot\cancel{7}} = \dfrac{2}{5}$

**13.** $\dfrac{17 \text{ hr}}{41 \text{ hr}} = \dfrac{17}{41}$

**15.** $\dfrac{\$121}{\$423} = \dfrac{121}{423}$

**17.** (a) $\dfrac{15}{35} = \dfrac{3\cdot\cancel{5}}{\cancel{5}\cdot 7} = \dfrac{3}{7}$

(b) $\dfrac{35}{15} = \dfrac{\cancel{5}\cdot 7}{\cancel{5}\cdot 3} = \dfrac{7}{3}$

**19.** (a) $\dfrac{\text{wins}}{\text{losses}} = \dfrac{23}{14}$ (b) $\dfrac{\text{losses}}{\text{wins}} = \dfrac{14}{23}$

**21.** (a) $\dfrac{8 \text{ grams}}{20 \text{ grams}} = \dfrac{4\cdot 2}{4\cdot 5} = \dfrac{2}{5}$

**23.** $\dfrac{3500 \text{ ft}}{4200 \text{ ft}} = \dfrac{5\cdot 7}{6\cdot 7} = \dfrac{5}{6}$

**25.** $\dfrac{315 \text{ cal}}{19 \text{ gram fat}} = 21\dfrac{11}{19}$ cal per gram fat

**27.** $\frac{315 \text{ mi}}{14 \text{ gal}} = 22\frac{1}{2} \text{ mi/gal}$

**29.** $\frac{\$304}{38 \text{ hr}} = \$8$ per hour

**31.** $\frac{320 \text{ mi}}{6 \text{ hr}} = 53\frac{1}{3}$ mph

**33.** $\frac{616 \text{ mi}}{28 \text{ gal}} = 22$ mi/gal

**35.** $\frac{\$108}{9 \text{ books}} = \$12$ per book

**37.** (a) $\frac{90}{5} = 18$ students per instructor

(b) $\frac{30}{2} = 15$ students per tutor

(c) $\frac{15}{1} = \frac{90}{n} \Rightarrow n = 6$ tutors

**39.** (a) $\frac{96}{8} = \$12$ per glass for box of 8

(b) $\frac{78}{6} = \$13$ per glass for box of 6

(c) $12 \div 8 = \$1.50$ per glass

$13 \div 6 = \$2.1\overline{6}$ per glass

Box of 8 is better buy.

**41.** (a) $\frac{\$32}{4 \text{ CDs}} = \$8$ per CD

$\frac{\$48}{6 \text{ CDs}} = \$8$ per CD

(b) Both offer the same unit price.

**43.** $\frac{3161}{31} \approx 102$ sales per day

**45.** $\frac{\$562,500}{4500 \text{ shares}} = \$125$ per share

**Cumulative Review**

**47.** $x - 12 = 25 \Rightarrow x = 25 + 12 = 37$

check: $37 - 12 \stackrel{?}{=} 25,\ 25 = 25$

**49.** $5x - 4x + 6 = 14 \Rightarrow x = 14 - 6 = 8$

check: $5(8) - 4(8) + 6 \stackrel{?}{=} 16,\ 16 = 16$

**51.** $101°\text{F} - (-8°\text{F}) = 109°\text{F}$

**4.6 Understanding the Concept**

1. A ratio is a fraction and reducing does not change the value of the fraction.

**4.6 Exercises**

1. When the numerator and denominator of a fraction are equal, the fraction is equal to 1. Thus both fractions are equal to 1.

3. Two fractions are equal if their cross products are equal.

5. $\frac{4}{9} = \frac{28}{63}$

7. $\frac{12 \text{ goals}}{7 \text{ games}} = \frac{24 \text{ goals}}{14 \text{ games}}$

9. $\frac{3}{8} = \frac{18}{48}$

11. $\frac{2 \text{ cups}}{50 \text{ grams}} = \frac{6 \text{ cups}}{150 \text{ grams}}$

**13.** $\frac{3\frac{1}{2}\text{ rotations}}{2\text{ min}} = \frac{14\text{ rotations}}{8\text{ min}}$

**15.** $\frac{4\text{ made}}{7\text{ attempts}} = \frac{12\text{ made}}{21\text{ attempts}}$

**17.** $7(32) = 224 = 8(28)$

**19.** $9(66) = 594 \neq 462 = 11(42)$

**21.** $2(28) = 56 = 7(8)$, yes

**23.** $13(29) = 377 \neq 494 = 19(26)$, no

**25.** $2(135) = 270 \neq 715 = 11(65)$, no

**27.** $\frac{x}{8} = \frac{5}{2} \Rightarrow 2x = 40 \Rightarrow x = 20$

check: $\frac{20}{8} \stackrel{?}{=} \frac{5}{2}, \ \frac{5}{2} = \frac{5}{2}$

**29.** $\frac{12}{x} = \frac{3}{5} \Rightarrow 3x = 60 \Rightarrow x = 20$

check: $\frac{12}{20} \stackrel{?}{=} \frac{3}{5}, \ \frac{3}{5} = \frac{3}{5}$

**31.** $\frac{12}{18} = \frac{x}{21} \Rightarrow 18x = 252 \Rightarrow x = 14$

check: $\frac{12}{18} \stackrel{?}{=} \frac{14}{21}, \ \frac{2}{3} = \frac{2}{3}$

**33.** $\frac{15}{6} = \frac{5}{x} \Rightarrow 15x = 30 \Rightarrow x = 2$

check: $\frac{15}{6} \stackrel{?}{=} \frac{5}{2}, \ \frac{5}{2} = \frac{5}{2}$

**35.** $\frac{80\text{ gal}}{24\text{ acres}} = \frac{20\text{ gal}}{n\text{ acres}}$

$80n = 480$

$n = 6$

**37.** $\frac{n\text{ gram}}{15\text{ liters}} = \frac{12\text{ gram}}{45\text{ liters}}$

$45n = 180$

$n = 4$

**39.** $\frac{400}{5} \stackrel{?}{=} \frac{675}{9}$

$400(9) = 3600 \neq 3375 = 5(675)$, no

**41.** $\frac{18}{30} \stackrel{?}{=} \frac{15}{25}$, $18(25) = 450 = 30(15)$, yes

**43.** $3(5) = 15 \neq 16 = 4(4)$, no

**45.** $\frac{2\text{ cups}}{50\text{ g}} = \frac{5\text{ cups}}{x} \Rightarrow 2x = 250, \ x = 125\text{ g}$

**47.** $\frac{25\text{ min}}{2\text{ rows}} = \frac{t}{12\text{ rows}}$

$2t = 300$

$t = 150\text{ min} = 2\frac{1}{2}\text{ hr}$

**49.** $\frac{1\text{ day}}{16\text{ million}} = \frac{7}{n}$

$n = 16(7) = 112\text{ million}$

**51.** $\frac{210\text{ ft}}{3\text{ in.}} = \frac{x}{5\text{ in.}}$

$3x = 1050$

$x = 350\text{ ft}$

**53.** $\dfrac{5 \text{ cups water}}{2 \text{ cups punch}} = \dfrac{x}{8 \text{ cups punch}}$
$2x = 40$
$x = 20$ cups water

**55.** $\dfrac{8 \text{ shares}}{5 \text{ held}} = \dfrac{n}{850 \text{ held}}$
$5n = 6800$
$n = 1360$ shares

**57.** $\dfrac{100 \text{ watt}}{30 \text{ mm}} = \dfrac{140 \text{ watt}}{t} \Rightarrow 100t = 4200$
$t = 42$ mm

**59.** $\dfrac{12 \text{ ft}}{18 \text{ ft}} = \dfrac{\text{W}}{30 \text{ ft}} \Rightarrow 18W = 360,\ W = 20$ ft

**61.** $\dfrac{6}{7} = \dfrac{\$2400}{K} \Rightarrow 6K = 16{,}800,\ K = \$2800$

**63.** $\dfrac{\frac{1}{3}}{\frac{1}{8}} = \dfrac{\frac{1}{4}}{\frac{3}{32}}$

**65.** $\dfrac{\$325}{\$1950} = \dfrac{\$40}{x} \Rightarrow 325x = 78{,}000$
$x = \$240$

**67.** $\dfrac{\$325}{\$1950} = \dfrac{\$32}{x} \Rightarrow 325x = 62{,}400$
$x = \$192$

**69.** $\dfrac{\$325}{\$1950} = \dfrac{\$215}{x} \Rightarrow 325x = 419{,}250$
$x = \$1290$

**71.** $\dfrac{L}{W} = \dfrac{2 \text{ in.}}{3 \text{ in.}} = \dfrac{6 \text{ in.}}{W} \Rightarrow 2W = 18$
$W = 9$ in.
$\dfrac{H}{L} = \dfrac{5 \text{ in.}}{2 \text{ in.}} = \dfrac{H}{6 \text{ in.}} \Rightarrow 2H = 30$
$H = 15$ in.

**73.** (a) $\dfrac{24}{4} = 6 = \dfrac{36}{6}$, yes
(b) $\dfrac{24}{3} = 8 \neq \dfrac{36}{5} = 7\dfrac{1}{5}$, no

## Cumulative Review

**75.** $2x + 6 = 28$

**77.** $x - 8 = 9$

**79.** $5(-1) + 50(5) + 10(-3) = 215$ points

## Putting Your Skills to Work, Problems for Individual Analysis

**1.** $\dfrac{3 \text{ rolls}}{1 \text{ person}} = \dfrac{n}{192 \text{ people}}$
$n = 3(192) = 576$ rolls

**2.** (a) $\dfrac{2 \text{ lb}}{24 \text{ people}} = \dfrac{x}{192 \text{ people}}$
$24x = 192(2) = 384 \Rightarrow x = 16$ lb ham
(b) $\dfrac{2 \text{ lb}}{16 \text{ people}} = \dfrac{x}{192 \text{ people}}$
$16x = 192(2) = 384 \Rightarrow x = 24$ lb turkey
(c) $\dfrac{2 \text{ lb}}{16 \text{ people}} = \dfrac{x}{192 \text{ people}}$
$16x = 192(2) = 384 \Rightarrow x = 24$ lb cheese

**3.** $\frac{1 \text{ basket}}{6 \text{ people}} = \frac{x}{192 \text{ people}}$
$6x = 192 \Rightarrow x = 32$ baskets

**4.** $\frac{1 \text{ cake}}{24 \text{ people}} = \frac{x}{192 \text{ people}}$
$24x = 192 \Rightarrow x = 8$ cakes

**5.** $\frac{3 \text{ parts nuts}}{2 \text{ parts candy}} = \frac{15 \text{ bags of nuts}}{b}$
$3b = 2(15) = 30 \Rightarrow b = 10$ bags candy

**6.** $\frac{6 \text{ yellow}}{9 \text{ white}} = \frac{72 \text{ yellow}}{w}$
$6w = 9(72) = 648$
$w = 108$ white daisies

**7.** $\frac{2 \text{ yd white}}{3 \text{ yd yellow}} = \frac{w}{30 \text{ yd yellow}}$
$3w = 2(30) = 60 \Rightarrow w = 20$ yd white

**8.** $\frac{4 \text{ cups flour}}{48 \text{ rolls}} = \frac{c}{576 \text{ rolls}}$
$48c = 4(576) = 2304 \Rightarrow c = 48$ cups flour

**9.** $\frac{2 \text{ cups milk}}{48 \text{ rolls}} = \frac{c}{576 \text{ rolls}}$
$48c = 2(576) = 1152 \Rightarrow c = 24$ cups milk

### Putting Your Skills to Work, Problems for Group Investigation and Study

**1.** $\frac{\$3}{48 \text{ rolls}} = \frac{c}{576} \Rightarrow 48c = 3(576) = 1728$
$c = \$36$

**2.** $6(16) + 5(24) + 4(24) = \$312$

**3.** $\frac{\$5}{8 \text{ baskets}}(32 \text{ baskets}) = \$20$

**4.** $\frac{\$28}{2 \text{ cakes}}(8 \text{ cakes}) = \$112$

**5.** (a)
$\frac{4 \text{ servings}}{\text{person}}(192 \text{ people}) = 768$ servings
$\frac{2 \text{ cups}}{24 \text{ servings}} = \frac{c}{768 \text{ servings}}$
$c = 64$ cups

(b) $\frac{4 \text{ cups}}{24 \text{ servings}} = \frac{c}{768 \text{ servings}}$
$24c = 3072 \Rightarrow c = 128$ cups

(c) $\frac{3 \text{ cups}}{24 \text{ servings}} = \frac{c}{768 \text{ servings}}$
$c = 96$ cups

**6.** $\frac{24 \text{ servings}}{\$3} = \frac{768 \text{ servings}}{c}$
$c = \$96$

**7.** $15(3) + 10(2) = \$65$

### Chapter 4 Review Problems

**1.** prime number – a whole number greater than 1 that is divisible only by itself and 1.

**2.** composite number – a whole number greater than 1 that is divisible by a whole number other than itself and 1.

**3.** proper fraction – a fraction that describes a quantity less than 1.

**4.** improper fraction – a fraction that describes a quantity greater than or equal to one.

**5.** mixed number – the sum of a whole number greater than zero and a proper fraction.

**6.** equivalent fractions – fractions that look different but have the same value.

**7.** ratio – a comparison of two quantities that have the same units.

**8.** rate – a ratio that compares different units.

**9.** proportion – two rates or ratios that are equal.

**10.** 322,970 ends in 0 and is thus divisible by 2 and 5. The sum of the digits is 23 which is not divisible by 3 and therefore 322,970 is not divisible by 3.

**11.** 41,592 is even and thus divisible by 2. 41,592 does not end in 0 or 5 and is thus not divisible by 5. The sum of the digits is 21 which is divisible by 3 and thus 41,592 is divisible by 3.

**12.** 0: neither; 7, 11: prime; 21, 50, 25, 51: composite

**13.** 1: neither; 7, 13, 41: prime; 32, 12, 50, 6: composite

**14.** $36 = 2 \cdot 2 \cdot 3 \cdot 3 = 2^2 \cdot 3^2$

**15.** $56 = 2 \cdot 2 \cdot 2 \cdot 7 = 2^3 \cdot 7$

**16.** $425 = 5 \cdot 5 \cdot 17 = 5^2 \cdot 17$

**17.** $90 = 2 \cdot 3 \cdot 3 \cdot 5 = 2 \cdot 3^2 \cdot 5$

**18.** $\frac{1}{0}$ undefined

**19.** $\frac{0}{6} = 0$

**20.** $\frac{z}{z} = 1$

**21.** $\frac{a}{a} = 1$

**22.** $\frac{7}{20}$

**23.** $\frac{27}{69} = \frac{\cancel{3} \cdot 9}{\cancel{3} \cdot 23} = \frac{9}{23}$

**24.** $\frac{42}{5} = 8\frac{2}{5}$

**25.** $\frac{55}{6} = 9\frac{1}{6}$

**26.** $\frac{56}{7} = 8$

**27.** $\frac{42}{7} = 6$

**28.** $2\frac{1}{3} = \frac{3 \cdot 2 + 1}{3} = \frac{7}{3}$

**29.** $4\frac{3}{5} = \frac{5 \cdot 4 + 3}{5} = \frac{23}{5}$

**30.** $10\frac{2}{5} = \frac{5 \cdot 10 + 2}{5} = \frac{52}{5}$

**31.** $11\frac{1}{4} = \frac{4 \cdot 11 + 1}{4} = \frac{45}{4}$

**32.** $\frac{1}{9} \cdot \frac{2}{2} = \frac{2}{18}$

**33.** $\frac{1}{2} \cdot \frac{11}{11} = \frac{11}{22}$

**34.** $\frac{2}{3}\cdot\frac{9}{9}=\frac{18}{27}$

**35.** $\frac{3}{4}\cdot\frac{9}{9}=\frac{27}{36}$

**36.** $\frac{4}{5}\cdot\frac{7x}{7x}=\frac{28x}{35x}$

**37.** $\frac{6}{11}\cdot\frac{3y}{3y}=\frac{18y}{33y}$

**38.** $\frac{55}{75}=\frac{\cancel{5}\cdot 11}{\cancel{5}\cdot 15}=\frac{11}{15}$

**39.** $\frac{48}{54}=\frac{\cancel{6}\cdot 8}{\cancel{6}\cdot 9}=\frac{8}{9}$

**40.** $\frac{108}{36}=\frac{\cancel{36}\cdot 3}{\cancel{36}}=3$

**41.** $\frac{175}{75}=\frac{\cancel{25}\cdot 7}{\cancel{25}\cdot 3}=\frac{7}{3}=2\frac{1}{3}$

**42.** $\frac{25\cancel{x}}{60\cancel{x}}=\frac{\cancel{5}\cdot 5}{\cancel{5}\cdot 12}=\frac{5}{12}$

**43.** $\frac{84\cancel{x}}{105\cancel{x}y}=\frac{\cancel{21}\cdot 4}{\cancel{21}\cdot 5y}=\frac{4}{5y}$

**44.** $\frac{-16}{18}=-\frac{\cancel{2}\cdot 8}{\cancel{2}\cdot 9}=-\frac{8}{9}$

**45.** $\frac{24}{-36}=-\frac{\cancel{12}\cdot 2}{\cancel{12}\cdot 3}=-\frac{2}{3}$

**46.** $\frac{y^5}{y^3}=y^{5-3}=y^2$

**47.** $\frac{3^2}{3^3}=3^{2-3}=3^{-1}=\frac{1}{3}$

**48.** $\frac{a^2b^4}{a^5b^5}=a^{2-5}b^{4-5}=a^{-3}b^{-1}=\frac{1}{a^3b}$

**49.** $\frac{x^5y^3}{x^2y^9}=x^{5-2}y^{3-9}=x^3y^{-6}=\frac{x^3}{y^6}$

**50.** $\frac{2^3x^0}{2^6x^9}=\frac{2^{3-6}\cdot 1}{x^9}=\frac{2^{-3}}{x^9}=\frac{1}{2^3x^9}$

**51.** $\frac{3^2y^0}{3^3y^6}=\frac{3^{2-3}\cdot 1}{y^6}=\frac{3^{-1}}{y^6}=\frac{1}{3y^6}$

**52.** $\frac{20x^5}{35x^9}=\frac{\cancel{5}\cdot 4x^{5-9}}{\cancel{5}\cdot 7}=\frac{4x^{-4}}{7}=\frac{4}{7x^4}$

**53.** $\frac{\overset{3}{\cancel{18}}\,y^6}{\cancel{6}\,y^4}=3y^{6-4}=3y^2$

**54.** $(3y^2)^3=3^3y^{2\cdot 3}=3^3y^6$

**55.** $(2^4x)^2=2^{4\cdot 2}x^2=2^8x^2$

**56.** $\left(\frac{3}{y}\right)^2=\frac{3^2}{y^2}=\frac{9}{y^2}$

**57.** $\left(\frac{x}{2}\right)^3=\frac{x^3}{2^3}=\frac{x^3}{8}$

**58.** $\frac{20}{46}=\frac{\cancel{2}\cdot 10}{\cancel{2}\cdot 23}=\frac{10}{23}$

**59.** $\frac{15}{25}=\frac{\cancel{5}\cdot 3}{\cancel{5}\cdot 5}=\frac{3}{5}$

**60.** $\dfrac{35\text{ yd}}{55\text{ yd}} = \dfrac{\cancel{5}\cdot 7}{\cancel{5}\cdot 11} = \dfrac{7}{11}$

**61.** $\dfrac{24}{28} = \dfrac{\cancel{4}\cdot 6}{\cancel{4}\cdot 7} = \dfrac{6}{7}$

**62.** $\dfrac{28}{24} = \dfrac{\cancel{4}\cdot 7}{\cancel{4}\cdot 6} = \dfrac{7}{6}$

**63.** $\dfrac{\$35}{7\text{ washcloths}} = \$5$ per washcloth

**64.** $\dfrac{112\text{ in.}}{28\text{ hr}} = 4$ in. per hr

**65.** $\dfrac{210\text{ mi}}{8\text{ hr}} = 26\dfrac{1}{4}$ mph

**66.** $\dfrac{10\text{ fat calories}}{2\text{ gram protein}}$
$= 5$ fat calories per gram protein

**67.** $\dfrac{110\text{ calories}}{24\text{ gram carbohydrates}}$
$= 4\dfrac{7}{12}$ calories per gram carbohydrates

**68.** (a) $\dfrac{32\text{ legal secretaries}}{16\text{ lawyers}}$
$= 2$ legal secretaries per lawyer

(b) $\dfrac{12\text{ paralegals}}{4\text{ lawyers}}$
$= 3$ paralegals per lawyer

(c) $\dfrac{12\text{ paralegals}}{4\text{ lawyers}} = \dfrac{p}{60\text{ lawyers}}$
$p = 180$ paralegals

**69.** (a) $\dfrac{\$72}{6\text{ CD's}} = \$12$ per CD
$\dfrac{\$96}{8\text{ CD's}} = \$12$ per CD

(b) Both have the same unit price.

**70.** (a) $\dfrac{24{,}000}{31{,}000} = \dfrac{24}{31}$

(b) $\dfrac{\$45{,}000}{30\text{ days}} = \$1500$ per day

**71.** $\dfrac{2}{7} = \dfrac{14}{49}$

**72.** $\dfrac{2\text{ teachers}}{50\text{ students}} = \dfrac{6\text{ teachers}}{150\text{ students}}$

**73.** $\dfrac{2\text{ in.}}{190\text{ mi}} = \dfrac{6\text{ in.}}{570\text{ mi}}$

**74.** $\dfrac{234\text{ mi}}{9\text{ gal}} = \dfrac{468\text{ mi}}{18\text{ gal}}$

**75.** $3(70) = 210 \neq 156 = 4(39),\ \dfrac{3}{4} \neq \dfrac{39}{70}$

**76.** $13(84) = 1092 = 91(12),\ \dfrac{13}{91} = \dfrac{12}{84}$

**77.** $3(20) = 60 = 5(12)$, yes

**78.** $4(46) = 184 \neq 140 = 7(20)$, no

**79.** $\dfrac{x}{8} = \dfrac{2}{15} \Rightarrow 15x = 8(2) = 16 \Rightarrow x = \dfrac{16}{15}$

check: $\dfrac{\frac{16}{15}}{8} \stackrel{?}{=} \dfrac{2}{15},\ \dfrac{2}{15} = \dfrac{2}{15}$

**80.** $\frac{6}{5}=\frac{54}{x},\ 6x=5(54)=270,\ x=45$

check: $\frac{6}{5}\stackrel{?}{=}\frac{54}{45},\ \frac{6}{5}=\frac{6}{5}$

**81.** $\frac{24}{40}=\frac{x}{5}\Rightarrow 40x=24(5)=120\Rightarrow x=3$

check: $\frac{24}{40}\stackrel{?}{=}\frac{3}{5},\ \frac{3}{5}=\frac{3}{5}$

**82.** $\frac{17\text{ quarts}}{47\text{ square feet}}=\frac{n\text{ quarts}}{94\text{ square feet}}$

$47n=17(94)=1598$

$n=34$

**83.** $\frac{10\text{ mi}}{2\text{ hr}}=\frac{25\text{ mi}}{n\text{ hr}}$

$10n=2(25)=50$

$n=5$

**84.** $\frac{120\text{ mi}}{1\text{ in.}}=\frac{x\text{ mi}}{3\text{ in.}}$

$x=120(3)=360\text{ mi}$

**85.** $\frac{75\text{ rpm}}{12\text{ mph}}=\frac{100\text{ rpm}}{r}$

$75r=1200$

$r=16\text{ mph}$

**86.** $\frac{4\text{ ft wide}}{7\text{ ft long}}=\frac{W}{14\text{ ft long}}\Rightarrow 7W=4(14)$

$7W=56\Rightarrow W=8\text{ ft}$

**87.** (a) $\frac{\$400}{\$340}=\frac{\$60}{x}\Rightarrow 400x=60(340)$

$400x=20{,}400\Rightarrow x=\$51$

(b) $\frac{400}{340}=\frac{20}{x}\Rightarrow 400x=6800\Rightarrow x=\$17$

## How Am I Doing? Chapter 4 Test

**1.** 230 is even and therefore divisible by 2. 230 ends in 0 and therefore divisible by 5. Since the sum of the digits, $2+3=5$, is not divisible by 3, 230 is not divisible by 3.

**2.** (a) $27=3^3$ : composite
(b) 1 : neither
(c) 19 : prime

**3.** $84=2\cdot 2\cdot 3\cdot 7=2^2\cdot 3\cdot 7$

**4.** $120=2^3\cdot 3\cdot 5$

**5.** $\frac{0}{4}=0$

**6.** $\frac{t}{t}=1$

**7.** $\frac{12}{0}$ undefined

**8.** $\frac{17}{36}$

**9.** (a) $\frac{16}{12+7+16}=\frac{16}{35}$

(b) $\frac{7+16}{12+7+16}=\frac{23}{35}$

**10.** $\frac{12}{3}=4$

**11.** $\frac{8}{5}=1\frac{3}{5}$

**12.** $7\frac{1}{6}=\frac{7\cdot 6+1}{6}=\frac{43}{6}$

**13.** $\frac{1}{3}\cdot\frac{4}{4}=\frac{4}{12}$

**14.** $\frac{7}{8}\cdot\frac{5}{5}=\frac{35}{40}$

**15.** $\frac{4}{9}\cdot\frac{3y}{3y}=\frac{12y}{27y}$

**16.** $-\frac{18}{56}=-\frac{\cancel{2}\cdot 9}{\cancel{2}\cdot 28}=-\frac{9}{28}$

**17.** $\frac{16x}{32x^2y}=\frac{\cancel{2}\cdot\cancel{2}\cdot\cancel{2}\cdot\cancel{2}\cancel{x}}{\cancel{2}\cdot\cancel{2}\cdot\cancel{2}\cdot\cancel{2}\cdot 2\cancel{x}xy}=\frac{1}{2xy}$

**18.** $\frac{8^2}{7^3}=\frac{8^2}{7^3}$

**19.** $\frac{y^3z^4}{y^7z}=\frac{z^{4-1}}{y^{7-3}}=\frac{z^3}{y^4}$

**20.** $\frac{42x^7y^6}{36x^0y^9}=\frac{6\cdot 7x^7}{6\cdot 6\cdot 1y^{9-6}}=\frac{7x^7}{6y^3}$

**21.** $(2y^4)^3=2^3y^{4\cdot 3}=2^3y^{12}$

**22.** $(x^3)^5=x^{3\cdot 5}=x^{15}$

**23.** $\left(\frac{x}{3}\right)^2=\frac{x^2}{3^2}=\frac{x^2}{9}$

**24.** $\frac{16}{28+16}=\frac{16}{44}=\frac{\cancel{4}\cdot 4}{\cancel{4}\cdot 11}=\frac{4}{11}$

**25.** $\frac{150\text{ calories}}{20\text{ min}}=7\frac{1}{2}$ cal per min

**26.** (a)

$$\frac{\$36}{12\text{ reams}}=\frac{\$3}{1\text{ reams}};\quad\frac{\$96}{48\text{ reams}}=\frac{\$2}{1\text{ reams}}$$

(b) \$2 per ream is the better deal.

**27.** $20(13)=260=52(5)$, yes

**28.** $\frac{4}{6}=\frac{20}{x}\Rightarrow 4x=120\Rightarrow x=30$

**29.** $\frac{1\text{ lb}}{1200\text{ ft}^2}=\frac{x}{6000\text{ ft}^2}$

$1200x=6000$

$x=5$ lb

**30.** $\frac{2\text{ tablespoons}}{400\text{ square ft}}\cdot\frac{4}{4}=\frac{8\text{ tablespoons}}{1600\text{ square ft}}$

You will need 8 tablespoons.

### Cumulative Test for Chapters 1-4

**1.** $-2+6+(-8)+1=4+(-7)=-3$

**2.** $5-9-7-3=-4-10=-14$

**3.** (a) $25-x=18$
(b) $x=25-18=7$

**4.** $10^3=10\cdot 10\cdot 10=1000$

**5.** $8^2=64$

**6.** $12^1=12$

**7.** $(3x)(5x^2)=3\cdot 5x^{1+2}=15x^3$

**8.** $2(x+3)=2x+2(3)=2x+6$

**9.** $4(y-2)=4y-4(2)=4y-8$

**10.** $a(b+c)=ab+ac$

**11.** $6n\big|_{n=3}=6(3)=18$

**12.** $-6n\big|_{n=-3}=-6(-3)=18$

**13.** $\angle a+48°=180° \Rightarrow \angle a=132°$

**14.** $A=LW=(2x^3+1)(8x)=16x^4+8x$

**15.** 440 is even and thus divisible by 2. Since the sum of the digits, $4+4=8$, is not divisible by 3, 440 is not divisible by 3. 440 ends in 0 and is thus divisible by 5.

**16.** $110=2\cdot5\cdot11$

**17.** $\frac{37}{4}=9\frac{1}{4}$

**18.** $\frac{40}{3}=13\frac{1}{3}$

**19.** $\frac{5}{12}\cdot\frac{5x}{5x}=\frac{25x}{60x}$

**20.** $\frac{90n^2}{54n}=\frac{1\cancel{8}\cdot5\cancel{n}n}{1\cancel{8}\cdot3\cancel{n}}=\frac{5n}{3}$

**21.** $\frac{8a^7}{32a^5}=\frac{\cancel{8}\cancel{a^5}\cdot a^2}{\cancel{8}\cdot4\cancel{a^5}}=\frac{a^2}{4}$

**22.** $\left(\frac{x}{3}\right)^3=\frac{x^3}{3^3}=\frac{x^3}{27}$

**23.** (a) $\frac{44}{16}=\frac{\cancel{4}\cdot11}{\cancel{4}\cdot4}=\frac{11}{4}$

(b) $\frac{16}{44}=\frac{\cancel{4}\cdot4}{\cancel{4}\cdot11}=\frac{4}{11}$

**24.** $\frac{310\text{ mi}}{6\text{ hr}}=51\frac{2}{3}\text{ mph}$

**25.** (a) $\frac{\$108}{12\text{ lessons}}=\$9\text{ per lesson}$

$\frac{\$150}{15\text{ lessons}}=\$10\text{ per lesson}$

(b) The 12 lesson special is the better deal.

**26.** $\frac{150}{4}=\frac{900}{x}$

$150x=3600$

$x=24$ defective parts

**27.** (a) $\frac{200}{350}=\frac{\cancel{50}\cdot4}{\cancel{50}\cdot7}=\frac{4}{7}$

(b) $\frac{350-200}{350}=\frac{150}{350}=\frac{\cancel{50}\cdot3}{\cancel{50}\cdot7}=\frac{3}{7}$

**28.** $\frac{225\text{ words}}{3\text{ min}}=75$ words per min

**29.** $\frac{4\text{ oz}}{9\text{ g}}=\frac{12\text{ oz}}{x}\Rightarrow 4x=108$

$x=27$ g carbohydrates

# Chapter 5

## 5.1 Exercises

**1.** To multiply fractions: multiply numerator times numerator and denominator times denominator.

**3.** To split $\frac{1}{4}$ into 6 equal parts, we divide because we want to split an amount into equal parts.

**5.** Answers may vary but any word problem that requires taking $\frac{1}{3}$ of 90 or uses repeated addition of $\frac{1}{3}$ is correct.

**7.** $\frac{1}{2}\cdot\frac{2}{7}=\frac{2}{\boxed{14}}$

**9.** $\frac{1}{4}\div\frac{3}{7}=\frac{1}{4}\cdot\frac{\boxed{7}}{\boxed{3}}=\frac{7}{12}$

**11.** $\frac{1}{3}\cdot\frac{\boxed{5}}{\boxed{6}}=\frac{5}{18}$

**13.** $\frac{5}{7}\div\frac{4}{3}=\frac{5}{7}\boxed{\cdot}\frac{\boxed{3}}{\boxed{4}}=\frac{15}{28}$

**15.** $\frac{1}{3}\cdot\frac{1}{5}=\frac{1\cdot 1}{3\cdot 5}=\frac{1}{15}$

**17.** $\frac{5}{\underset{3}{\cancel{21}}}\cdot\frac{\overset{1}{\cancel{7}}}{8}=\frac{5\cdot 1}{3\cdot 8}=\frac{5}{24}$

**19.** $\frac{\overset{1}{\cancel{7}}}{\underset{3}{\cancel{12}}}\cdot\frac{\overset{2}{\cancel{8}}}{\underset{4}{\cancel{28}}}=\frac{1\cdot\overset{1}{\cancel{2}}}{3\cdot\underset{2}{\cancel{4}}}=\frac{1}{6}$

**21.** $\frac{\overset{1}{\cancel{3}}}{\underset{7}{\cancel{42}}}\cdot\frac{\overset{1}{\cancel{6}}}{\underset{5}{\cancel{15}}}=\frac{1\cdot 1}{7\cdot 5}=\frac{1}{35}$

**23.** $\frac{\overset{1}{\cancel{-3}}}{\underset{1}{\cancel{8}}}\cdot\frac{\overset{4}{\cancel{32}}}{\underset{2}{\cancel{-6}}}=\frac{\overset{2}{\cancel{4}}}{\underset{1}{\cancel{2}}}=2$

**25.** $\frac{\overset{4}{\cancel{16}}}{11}\cdot\frac{-18}{\underset{9}{\cancel{36}}}=-\frac{4\cdot\overset{2}{\cancel{18}}}{11\cdot\underset{1}{\cancel{9}}}=-\frac{8}{11}$

**27.** $\frac{-1}{63}\cdot\frac{-14}{18}=\frac{1}{9\cdot\cancel{7}}\cdot\frac{\cancel{7}\cdot\cancel{2}}{\cancel{2}\cdot 9}=\frac{1}{81}$

**29.** $-\overset{1}{\cancel{14}}\cdot\frac{1}{\underset{2}{\cancel{28}}}=-\frac{1}{2}$

**31.** $\frac{6}{\underset{7}{\cancel{35}}}\cdot\overset{1}{\cancel{5}}=\frac{6}{7}$

**33.** $\frac{2x}{\cancel{3}}\cdot\frac{\cancel{3}x}{5}=\frac{2x^2}{5}$

**35.** $\frac{6x^4}{\underset{1}{\cancel{7}}}\cdot\overset{2}{\cancel{14}}x^3=12x^7$

**37.** $\overset{4}{\cancel{8}}x^2\cdot\frac{3x^3}{\cancel{2}}=12x^5$

**39.** $\frac{\overset{1}{\cancel{2}}}{\underset{3}{\cancel{9}}}\cdot\frac{\overset{1}{\cancel{3}}}{\underset{4}{\cancel{8}}}=\frac{1}{12}$

**41.** $\frac{\overset{1}{\cancel{6}}\cancel{x}}{\underset{5}{\cancel{25}}}\cdot\frac{\overset{3}{\cancel{15}}}{\underset{2}{\cancel{12}}x^{\cancel{2}}}=\frac{3}{10x}$

**43.** $\frac{-\overset{1}{\cancel{3}}y^{\cancel{3}}}{\underset{5}{\cancel{20}}}\cdot\frac{\overset{3}{\cancel{12}}}{\underset{7}{\cancel{21}}\cancel{y^{2}}}=-\frac{3y}{35}$

**45.** $\frac{\overset{1}{\cancel{3}}x^2}{\underset{5}{\cancel{15}}}\cdot\frac{\overset{9}{\cancel{18}}x^3}{\underset{10}{\cancel{20}}}=\frac{9x^5}{50}$

**47.** $A=\frac{1}{2}bh=\frac{1}{2}(14)(9)=63\text{ m}^2$

**49.** $A=\frac{1}{2}bh=\frac{1}{2}(21)(40)=420\text{ in.}^2$

**51.** The reciprocal of $\frac{1}{8}$ is $\frac{8}{1}=8$

**53.** The reciprocal of 8 is $\frac{1}{8}$

**55.** The reciprocal of $\frac{2}{-5}$ is $\frac{-5}{2}=-\frac{5}{2}$

**57.** The reciprocal of $\frac{-x}{y}$ is $\frac{y}{-x}=-\frac{y}{x}$

**59.** $\frac{6}{14}\div\frac{3}{8}=\frac{\overset{2}{\cancel{6}}}{\underset{7}{\cancel{14}}}\cdot\frac{8}{\underset{1}{\cancel{3}}}=\frac{8}{7}$

**61.** $\frac{7}{24}\div\frac{9}{8}=\frac{7}{\underset{3}{\cancel{24}}}\cdot\frac{\cancel{8}}{9}=\frac{7}{27}$

**63.** $\frac{-1}{12}\div\frac{3}{4}=-\frac{1}{\underset{3}{\cancel{12}}}\cdot\frac{\cancel{4}}{3}=-\frac{1}{9}$

**65.** $\frac{-7}{24}\div\frac{9}{-8}=\frac{7}{\underset{3}{\cancel{24}}}\cdot\frac{\cancel{8}}{9}=\frac{7}{27}$

**67.** $\frac{8x^6}{15}\div\frac{16x^2}{5}=\frac{\cancel{8}x^{\cancel{6}4}}{\underset{3}{\cancel{15}}}\cdot\frac{\cancel{5}}{\underset{2}{\cancel{16}}\cancel{x^{2}}}=\frac{x^4}{6}$

**69.** $\frac{7x^4}{12}\div\frac{-28}{36x^2}=-\frac{\cancel{7}x^4}{\cancel{12}}\cdot\frac{\overset{3}{\cancel{36}}x^2}{\underset{4}{\cancel{28}}}=-\frac{3x^6}{4}$

**71.** $15\div\frac{3}{7}=\overset{5}{\cancel{15}}\cdot\frac{7}{\cancel{3}}=35$

**73.** $\frac{7}{22}\div 14=\frac{\cancel{7}}{22}\cdot\frac{1}{\underset{2}{\cancel{14}}}=\frac{1}{44}$

**75.** $14x^4\div\frac{7x^2}{3}=14x^4\cdot\frac{3}{7x^2}=6x^2$

**77.** $22x^3\div\frac{11}{6x^5}=22x^3\cdot\frac{6x^5}{11}=12x^8$

**79.** (a) $\frac{1}{15}\cdot\frac{25}{21}=\frac{5\cdot 5}{5\cdot 3\cdot 21}=\frac{5}{63}$

(b) $\frac{1}{15}\div\frac{25}{21}=\frac{1}{15}\cdot\frac{21}{25}=\frac{1}{3\cdot 5}\cdot\frac{3\cdot 7}{25}=\frac{7}{125}$

**81.** (a) $\frac{2x^2}{3} \div \frac{12}{21x^5} = \frac{2x^2}{3} \cdot \frac{21x^5}{12}$
$= \frac{2x^2}{3} \cdot \frac{3 \cdot 7x^5}{2 \cdot 6}$
$= \frac{7x^7}{6}$

(b) $\frac{2x^3}{3} \cdot \frac{12}{21x^5} = \frac{2x^3}{3} \cdot \frac{3 \cdot 4}{21x^5}$
$= \frac{8}{21x^3}$

**83.** $\frac{5x^7}{-27} \cdot \frac{-9}{20x^4} = \frac{5x^7}{3 \cdot 9} \cdot \frac{9}{5 \cdot 4x^4} = \frac{x^3}{12}$

**85.** $\frac{12x^6}{35} \div \frac{-16}{25x^2} = -\frac{3 \cdot 4x^6}{5 \cdot 7} \cdot \frac{5 \cdot 5x^2}{4 \cdot 4} = -\frac{15x^8}{28}$

**87.** $\frac{3}{12} \cdot 3300 = \$825$

**89.** $12 \div \frac{3}{4} = \overset{4}{\cancel{12}} \cdot \frac{4}{\cancel{3}} = 16$ pipes

**91.** $32 \text{ laps} \cdot \frac{\frac{1}{4} \text{ mi}}{\text{laps}} = \frac{32}{4} \text{ mi} = 8 \text{ mi}$

**93.** $\frac{120 \text{ qt}}{\text{vat}} \cdot \frac{\text{bottle}}{\frac{3}{4} \text{ qt}} = 120 \cdot \frac{4}{3} = 160 \frac{\text{bottles}}{\text{vat}}$

**95.** $\frac{5}{14} \div \frac{2}{21} \div \frac{\overset{5}{\cancel{15}}}{-\cancel{3}} = -\frac{\cancel{5}}{\underset{2}{\cancel{14}}} \cdot \frac{\overset{3}{\cancel{21}}}{2} \cdot \frac{1}{\cancel{5}} = -\frac{3}{4}$

**97.** $\frac{\frac{3}{8} \text{ pizza}}{\text{guest}} \cdot 17 \text{ guests} = 6\frac{3}{8} \to 7$ pizzas

**99.** $\frac{3}{4} \cdot \frac{x}{27} = \frac{4}{9}$, $x = \frac{4}{9} \cdot 36 = 16$

### Cumulative Review

**101.** $\frac{2}{3} = \frac{?}{15} \Rightarrow \frac{2}{3} \cdot \frac{5}{5} = \frac{10}{15}$

**103.** $120 = 2 \cdot 60 = 2 \cdot 2 \cdot 30 = 2 \cdot 2 \cdot 2 \cdot 15$
$= 2 \cdot 2 \cdot 2 \cdot 3 \cdot 5 = 2^3 \cdot 3 \cdot 5$

### 5.2 Exercises

**1.** Because $3 \cdot 4 = 12$ and there is no whole number that we can multiply by 5 to get 12.

**3.** (a) 6, 12, 18, 24; 8, 16, 24, 32
(b) 24

**5.** (a) 2, 4, 6, 8, 10; 5, 10, 15, 20, 25
(b) 10

**7.** (a) $12x$, $24x$, $36x$, $48x$
$18x$, $36x$, $54x$, $72x$
(b) $36x$

**9.** 2, 3

**11.** 7, $x$

**13.** $9 = 3 \cdot 3$, $18 = 2 \cdot 3 \cdot 3$
$\text{LCM} = 2 \cdot 3^2 = 18$

**15.** $16 = 2 \cdot 2 \cdot 2 \cdot 2$, $28 = 2 \cdot 2 \cdot 7$
$\text{LCM} = 2^4 \cdot 7 = 112$

**17.** $15 = 3 \cdot 5$, $21 = 3 \cdot 7$
$\text{LCM} = 3 \cdot 5 \cdot 7 = 105$

**19.** $40 = 2 \cdot 2 \cdot 2 \cdot 5,\ 60 = 2 \cdot 2 \cdot 3 \cdot 5$
$\text{LCM} = 2^3 \cdot 3 \cdot 5 = 120$

**21.** $5 = 5,\ 8 = 2^3,\ 12 = 2^2 \cdot 3$
$\text{LCM} = 2^3 \cdot 3 \cdot 5 = 120$

**23.** $7 = 7,\ 14 = 2 \cdot 7,\ 35 = 5 \cdot 7$
$\text{LCM} = 2 \cdot 5 \cdot 7 = 70$

**25.** $4x = 2^2 x,\ 7x = 7x,\ \text{LCM} = 4 \cdot 7x = 28x$

**27.** $21a = 3 \cdot 7a,\ 81a = 3^4 a$
$\text{LCM} = 3^4 \cdot 7a = 567a$

**29.** $18x = 2 \cdot 3^2 x,\ 45x^2 = 3^2 \cdot 5x^2$
$\text{LCM} = 2 \cdot 3^2 \cdot 5x^2 = 90x^2$

**31.** $22x^2 = 2 \cdot 11x^2,\ 4x^3 = 2^2 x^3$
$\text{LCM} = 2^2 \cdot 11x^3 = 44x^3$

**33.** $12x^2 = 2 \cdot 2 \cdot 3x^2,\ 5x = 5x,\ 3x^3 = 3x^3$
$\text{LCM} = 2 \cdot 2 \cdot 3 \cdot 5x^3 = 60x^3$

**35.** $12x = 2^2 \cdot 3x,\ 26 = 2 \cdot 13$
$156x^2 = 2^2 \cdot 3 \cdot 13x^2$
$\text{LCM} = 2^2 \cdot 3 \cdot 13x^2 = 156x^2$

**37.** $\dfrac{1 \text{ lap}}{4 \text{ min}} \cdot t = \dfrac{1 \text{ lap}}{6 \text{ min}} \cdot t + 1 \text{ lap}$
$\left(\dfrac{1}{4} - \dfrac{1}{6}\right)t = \dfrac{t}{12} = 1 \Rightarrow t = 12 \text{ min}$

**39.** $\dfrac{1 \text{ carton}}{6 \text{ min}} \cdot t = \dfrac{1 \text{ carton}}{8 \text{ min}} \cdot t + 1 \text{ carton}$
$\left(\dfrac{1}{6} - \dfrac{1}{8}\right) \cdot t = \dfrac{t}{24} = 1 \Rightarrow t = 24 \text{ min}$

**41.** There 35 min and 45 min between events on the two fields, respectively.
$35 = 5 \cdot 7, 45 = 5 \cdot 9, \text{LCM} = 5 \cdot 7 \cdot 9 = 315$
$315 \text{ min} = 5\dfrac{1}{4} \text{ hr} \to 1\text{:}15 \text{ P.M.}$

**43.** $2x^3 = 2x^3$
$8xy^2 = 2^3 xy^2$
$10x^2 y = 2 \cdot 5x^2 y$
$\text{LCM} = 2^3 \cdot 5 \cdot x^3 \cdot y^2$
$= 40x^3 y^2$

**45.** $2z^2 = 2z^2$
$5xyz = 5xyz$
$15xy = 3 \cdot 5xy$
$\text{LCM} = 2 \cdot 3 \cdot 5xyz^2$
$= 30xyz^2$

## Cumulative Review

**47.** $\dfrac{15}{2} = 7\dfrac{1}{2}$

**49.** $2 + 6(-1) \div 3 = 2 + (-6) \div 3$
$= 2 + (-2)$
$= 0$

**51.** $-11 + 14 + 7 + (-16) = -6$
The market fell by 6 points.

## 5.3 Exercises

**1.** When we add two fractions with the same denominator, we add the numerators, and the denominator stays the same.

**3.** $\frac{4}{5}+\frac{5}{9} \neq \frac{4+5}{5+9}=\frac{9}{14}$ No, we must find a common denominator when we add or subtract fractions with different denominators. Also, we do not add or subtract denominators.

**5.** $\frac{\boxed{2}}{7}+\frac{1}{7}=\frac{3}{7}$

**7.** $\frac{\boxed{2}}{4}-\frac{1}{4}=\frac{1}{4}$

**9.** $\frac{4}{21}+\frac{7}{21}=\frac{4+7}{21}=\frac{11}{21}$

**11.** $\frac{6}{17}-\frac{3}{17}=\frac{6-3}{17}=\frac{3}{17}$

**13.** $\frac{-13}{28}+\left(\frac{-11}{28}\right)=\frac{-13+(-11)}{28}=\frac{-24}{28}=-\frac{6}{7}$

**15.** $\frac{-31}{51}+\frac{11}{51}=\frac{-31+11}{51}=-\frac{20}{51}$

**17.** $\frac{7}{x}-\frac{5}{x}=\frac{7-5}{x}=\frac{2}{x}$

**19.** $\frac{31}{a}+\frac{8}{a}=\frac{31+8}{a}=\frac{39}{a}$

**21.** $\frac{x}{7}-\frac{5}{7}=\frac{x-5}{7}$

**23.** $\frac{y}{9}+\frac{14}{9}=\frac{y+14}{9}$

**25.** $4=2^2,\ 7=7,\ \text{LCD}=2^2\cdot 7=28$

**27.** $12=2^2\cdot 3,\ 15=3\cdot 5,\ \text{LCD}=2^2\cdot 3\cdot 5=60$

**29.** $\frac{1}{4}\cdot\frac{15}{15}=\frac{15}{60}$

**31.** $\frac{5}{6}\cdot\frac{10}{10}=\frac{50}{60}$

**33.** $\frac{11}{15}\cdot\frac{3}{3}-\frac{31}{45}=\frac{33-31}{45}=\frac{2}{45}$

**35.** $\frac{16}{24}-\frac{1}{6}\cdot\frac{4}{4}=\frac{16-4}{24}=\frac{12}{24}=\frac{1}{2}$

**37.** $\frac{3}{8}\cdot\frac{7}{7}+\frac{4}{7}\cdot\frac{8}{8}=\frac{21+32}{56}=\frac{53}{56}$

**39.** $\frac{-3}{4}\cdot\frac{5}{5}+\frac{1}{5}\cdot\frac{4}{4}=\frac{-15+4}{20}=-\frac{11}{20}$

**41.** $\frac{-2}{13}\cdot\frac{2}{2}+\frac{7}{26}=\frac{-4+7}{26}=\frac{3}{26}$

**43.** $\frac{-3}{14}\cdot\frac{5}{5}+\frac{-1}{10}\cdot\frac{7}{7}=\frac{-15+(-7)}{70}=-\frac{11}{35}$

**45.** $\frac{7}{10}\cdot\frac{7}{7}-\frac{5}{14}\cdot\frac{5}{5}=\frac{49-25}{70}=\frac{24}{70}=\frac{12}{35}$

**47.** $\frac{11}{15}\cdot\frac{4}{4}-\frac{5}{12}\cdot\frac{5}{5}=\frac{44-25}{60}=\frac{19}{60}$

**49.** $\frac{5}{2x}+\frac{8}{x}\cdot\frac{2}{2}=\frac{5+16}{2x}=\frac{21}{2x}$

**51.** $\frac{2}{7x}+\frac{3}{x}\cdot\frac{7}{7}=\frac{2+21}{7x}=\frac{23}{7x}$

**53.** $\frac{3}{2x}\cdot\frac{3}{3}+\frac{5}{6x}\cdot=\frac{9+5}{6x}=\frac{14}{6x}=\frac{7}{3x}$

**55.** $\frac{3}{x}\cdot\frac{y}{y}+\frac{4}{y}\cdot\frac{x}{x}=\frac{3y+4x}{xy}$

**57.** $\frac{2}{x}\cdot\frac{y}{y}-\frac{7}{y}\cdot\frac{x}{x}=\frac{2y-7x}{xy}$

**59.** $\frac{4x}{15}+\frac{3x}{5}\cdot\frac{3}{3}=\frac{4x+9x}{15}=\frac{13x}{15}$

**61.** $\frac{-3x}{10}\cdot\frac{2}{2}-\frac{7x}{20}=\frac{-6x-7x}{20}=\frac{-13x}{20}$

**63.** $\frac{x}{3}\cdot\frac{4}{4}+\frac{-11x}{12}=\frac{4x-11x}{12}=-\frac{7x}{12}$

**65.** $\frac{5}{6}\cdot\frac{5}{5}+\frac{4}{5}\cdot\frac{6}{6}=\frac{25+24}{30}=\frac{49}{30}=1\frac{19}{30}$

**67.** $\frac{3}{16}\cdot\frac{5}{5}+\left(\frac{-9}{20}\right)\cdot\frac{4}{4}=\frac{15+(-36)}{80}=-\frac{21}{80}$

**69.** $\frac{9}{y}\cdot\frac{x}{x}+\frac{1}{x}\cdot\frac{y}{y}=\frac{9x+y}{xy}$

**71.** $\frac{2x}{15}\cdot\frac{4}{4}+\frac{3x}{20}\cdot\frac{3}{3}=\frac{8x+9x}{60}=\frac{17x}{60}$

**73.** $\frac{2}{3}\cdot\frac{4}{4}+\frac{1}{4}\cdot\frac{3}{3}=\frac{8+3}{12}=\frac{11}{12}$ lb

**75.** (a) $\frac{3}{4}+\frac{1}{2}\cdot\frac{2}{2}=\frac{3+2}{4}=\frac{5}{4}=1\frac{1}{4}$ cups sugar

(b) $\frac{3}{4}-\frac{1}{2}\cdot\frac{2}{2}=\frac{3-2}{4}=\frac{1}{4}$ cup

**77.** $\frac{1}{3}\cdot\frac{4}{4}-\frac{1}{4}\cdot\frac{3}{3}=\frac{4-3}{12}=\frac{1}{12}$

**79.** (a) $\frac{1}{8}\cdot\frac{3}{3}+\frac{1}{12}\cdot\frac{2}{2}=\frac{3+2}{24}=\frac{5}{24}$

(b) $\frac{1}{8}\cdot\frac{3}{3}-\frac{1}{12}\cdot\frac{2}{2}=\frac{3-2}{24}=\frac{1}{24}$

**81.** $\frac{5}{30}\cdot\frac{4}{4}+\frac{3}{40}\cdot\frac{3}{3}+\frac{1}{8}\cdot\frac{15}{15}$

$=\frac{20+9+15}{120}=\frac{44}{120}=\frac{11}{30}$

**83.** $\frac{1}{3}\cdot\frac{4}{4}+\frac{1}{12}-\frac{1}{6}\cdot\frac{2}{2}=\frac{4+1-2}{12}=\frac{3}{12}=\frac{1}{4}$

## Cumulative Review

**85.** $x-3=70 \Rightarrow x=70+3=73$

**87.** $x-9=29 \Rightarrow x=29+9=38$

**89.** $3033-1295-469-387-287=\$595$

## 5.4 Understanding the Concept

(a)
$$\begin{array}{r} 25\frac{9}{12}\cdot\frac{2}{2} \\ +32\frac{15}{24} \\ \hline 57\frac{33}{24}=58\frac{9}{24} \end{array}$$

(b)
$$\begin{array}{r} 25\frac{9}{12}\rightarrow\frac{309}{12}\cdot\frac{2}{2} \\ +32\frac{15}{24}\rightarrow\frac{783}{24} \\ \hline \frac{1401}{24}=58\frac{9}{24} \end{array}$$

With large numbers it is easier to keep numbers as mixed numbers.

**Putting Your Skills to Work, Problems for Individual Analysis**

**1.** $\dfrac{1 \text{ beat}}{\text{quarter-note}} \cdot n \text{ quarter-note} = 4 \text{ beats}$

$n = 4$. Four quarter-notes

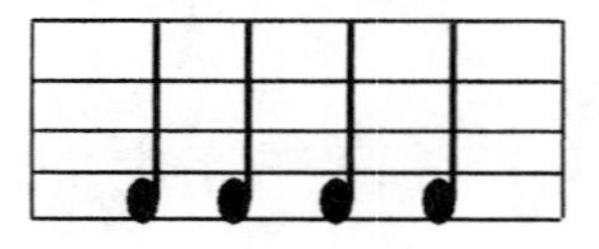

**2.** $\dfrac{2 \text{ beat}}{\text{half-note}} \cdot n \text{ half-note} = 4 \text{ beats}, n = 2$

Two half-notes

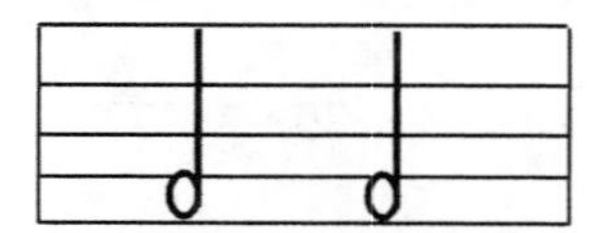

**3.** $\dfrac{2 \text{ beats}}{\text{half-note}} \cdot 1 + \dfrac{1 \text{ beat}}{\text{quarter-note}} \cdot 2 = 4 \text{ beats}$

One half-note and two quarter-notes

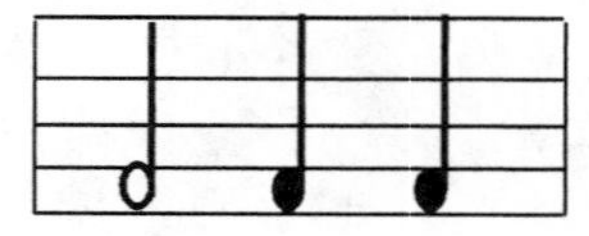

**4.** Four, since this measure contains two beats, i.e., two quarter-notes with 2(2) eighth-notes.

**5.** Four. Since the half-note takes up two beats out of six, there are four beats left for quarter-notes.

**6.** 1 half-note, 2 eighth-notes, and 1 quarter-note:

1 half-note and 4 eighth-notes:

1 half-note and two quarter-notes:

2 half-notes:

**Putting Your Skills to Work, Problems for Group Investigation and Study**

**1.** Think of $\frac{3}{4}$ as $3/4$. The numerator tells us how many quarter-notes can occur in a measure of 3 over 4. Since there are 3 beats for the measure and the quarter-note gets one beat. there are 3 quarter-notes in the measure.

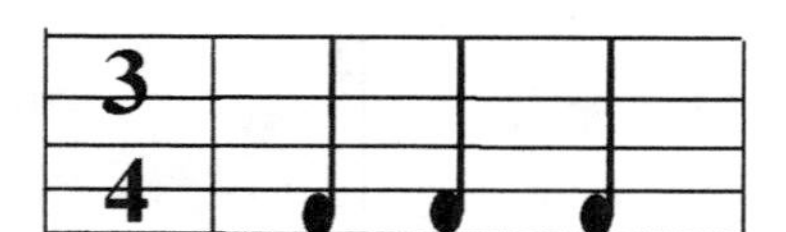


**2.** Think $3/4 = 6/8$. The numerator 6 indicates there are 6 eighth-notes.

**3.** Think $4/4 = 8/8$. There are 8 eighth-notes.

**4.** Two, think $4/4 = 2/2$. The numerator, 2, indicates that there are 2 half-notes.

**5.** 1 half-note and 2 eighth-notes:

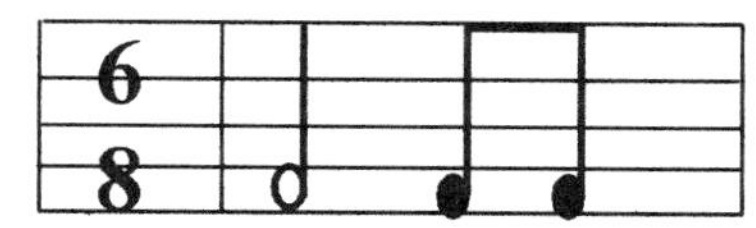


1 half-note and 1 quarter-note:

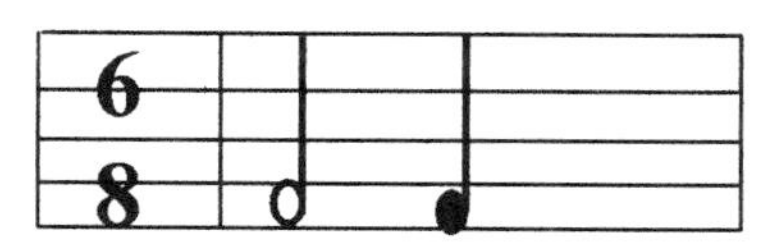


**6.** Two dotted half-notes would fill the measure:

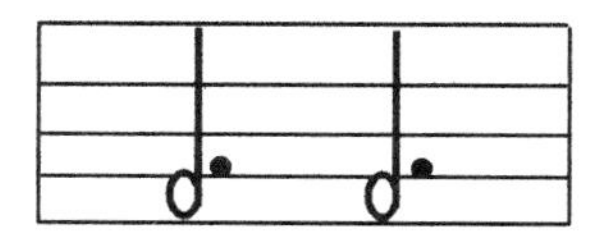

One dotted half-note and three quarter-notes:

## 5.4 Exercises

**1.** (a) Marcy did not change the mixed numbers to improper fractions before she multiplied.

(b) $2\frac{2}{3} \cdot 3\frac{4}{5} = \frac{8}{3} \cdot \frac{19}{5} = \frac{152}{15} = 10\frac{2}{15}$

**3.**
$$\begin{array}{r} 12\frac{3}{7} \\ +15\frac{2}{7} \\ \hline 27\frac{5}{7} \end{array}$$

**5.**
$$\begin{array}{r} 5\frac{5}{8} \\ +11\frac{1}{8} \\ \hline 16\frac{6}{8} = 16\frac{3}{4} \end{array}$$

**7.**
$$\begin{array}{rr} 11\frac{2}{3} \cdot \frac{4}{4} = & 11\frac{8}{12} \\ +7\frac{1}{4} \cdot \frac{3}{3} = & 7\frac{3}{12} \\ \hline & 18\frac{11}{12} \end{array}$$

**9.**
$$\begin{array}{rr} 14\frac{7}{9} \quad 14\frac{1}{4} \cdot \frac{3}{3} = & 14\frac{3}{12} \\ +6\frac{1}{3} \cdot \frac{4}{4} = & +6\frac{4}{12} \\ \hline & 20\frac{7}{12} \end{array}$$

**11.**
$$\begin{array}{rl} 6\frac{5}{6}\cdot\frac{4}{4} = & 6\frac{20}{24} \\ +4\frac{3}{8}\cdot\frac{3}{3} = & 4\frac{9}{24} \\ \hline & 10\frac{29}{24} = 11\frac{5}{24} \end{array}$$

**13.**
$$\begin{array}{r} 7\frac{4}{5} \\ -2\frac{1}{5} \\ \hline 5\frac{3}{5} \end{array}$$

**15.**
$$\begin{array}{rl} 9\frac{2}{3}\cdot\frac{2}{2} = 9\frac{4}{6} \\ -6\frac{1}{6} \quad -6\frac{1}{6} \\ \hline 3\frac{3}{6} = 3\frac{1}{2} \end{array}$$

**17.**
$$\begin{array}{rl} 11\frac{1}{5} = 10\frac{6}{25} \\ -6\frac{3}{5} \quad -6\frac{3}{5} \\ \hline 4\frac{3}{5} \end{array}$$

**19.**
$$\begin{array}{rl} 10\frac{5}{12}\cdot\frac{5}{5} = 10\frac{25}{60} = & 9\frac{85}{60} \\ -3\frac{9}{10}\cdot\frac{6}{6} = -3\frac{54}{60} = & -3\frac{54}{60} \\ \hline & 6\frac{31}{60} \end{array}$$

**21.**
$$\begin{array}{rl} 9 = & 8\frac{4}{4} \\ -2\frac{1}{4} & -2\frac{1}{4} \\ \hline & 6\frac{3}{4} \end{array}$$

**23.**
$$\begin{array}{rl} 25\frac{2}{3}\cdot\frac{7}{7} = & 25\frac{14}{21} \\ -6\frac{1}{7}\cdot\frac{3}{3} = & -6\frac{3}{21} \\ \hline & 19\frac{11}{21} \end{array}$$

**25.**
$$\begin{array}{rl} 1\frac{1}{6}\cdot\frac{4}{4} = & 1\frac{4}{24} \\ +\frac{3}{8}\cdot\frac{3}{3} = & +\frac{9}{24} \\ \hline & 1\frac{13}{24} \end{array}$$

**27.**
$$\begin{array}{rl} 8\frac{1}{4}\cdot\frac{3}{3} = & 8\frac{3}{12} \\ +3\frac{5}{6}\cdot\frac{2}{2} = & +3\frac{10}{12} \\ \hline & 11\frac{13}{12} = 12\frac{1}{12} \end{array}$$

**29.**
$$\begin{array}{rl} 32 = & 31\frac{9}{9} \\ -1\frac{2}{9} & -1\frac{2}{9} \\ \hline & 30\frac{7}{9} \end{array}$$

**31.** $2\frac{1}{5}\cdot 1\frac{2}{3} = \frac{11}{\not{5}}\cdot\frac{\not{5}}{3} = \frac{11}{3} = 3\frac{2}{3}$

**33.** $4\frac{1}{3}\cdot 2\frac{1}{4}=\frac{13}{\cancel{3}}\cdot\frac{\overset{3}{\cancel{9}}}{4}=\frac{39}{4}=9\frac{3}{4}$

**35.** $-\frac{3}{4}\cdot 9\frac{5}{7}=-\frac{3}{\cancel{4}}\cdot\frac{\overset{17}{\cancel{68}}}{7}=-\frac{51}{7}=-7\frac{2}{7}$

**37.** $2\frac{1}{4}\div(-4)=\frac{9}{4}\cdot\frac{1}{-4}=-\frac{9}{16}$

**39.** $4\frac{1}{2}\div 2\frac{1}{4}=\frac{\cancel{9}}{\cancel{2}}\cdot\frac{\overset{2}{\cancel{4}}}{\cancel{9}}=2$

**41.** $3\frac{1}{4}\div\frac{3}{8}=\frac{13}{\cancel{4}}\cdot\frac{\overset{2}{\cancel{8}}}{3}=\frac{26}{3}=8\frac{2}{3}$

**43.** $-5\div\frac{1}{4}=-5\cdot\frac{4}{1}=-20$

**45.** $1\frac{1}{4}\cdot 3\frac{2}{3}=\frac{5}{4}\cdot\frac{11}{3}=\frac{55}{12}=4\frac{7}{12}$

**47.** $6\frac{1}{2}\div\frac{3}{4}=\frac{13}{\cancel{2}}\cdot\frac{\overset{2}{\cancel{4}}}{3}=\frac{26}{3}=8\frac{2}{3}$

**49.** $7\frac{1}{2}\div(-8)=\frac{15}{2}\cdot\frac{1}{-8}=-\frac{15}{16}$

**51.** $26\div 6\frac{1}{2}=26\cdot\frac{2}{13}=4$ pieces

**53.** $\frac{1}{3}\cdot 7\frac{1}{5}=\frac{1}{3}\cdot\frac{36}{5}=\frac{12}{5}=2\frac{2}{5}$ ft

**55.** $2\frac{1}{4}\cdot 2=\frac{9}{4}\cdot 2=\frac{9}{2}=4\frac{1}{2}$ cups flour

**57.** $2\frac{1}{2}\cdot 4=\frac{5}{2}\cdot 4=10$ cups chocolate chips

**59.** $9\frac{1}{8}+12\frac{1}{3}+17\frac{1}{6}=9\frac{3}{24}+12\frac{8}{24}+17\frac{4}{24}$
$=38\frac{15}{24}=38\frac{5}{8}$ miles

**Cumulative Review**

**61.** $2+9\cdot 8=2+72=74$

**63.** $\frac{(5+7)}{(2\cdot 3)}=\frac{12}{6}=2$

**63.** $19(7)+28(2)-2(3)-1(5)=\$178$ Jan.
$26(7)+36(2)-3(3)-2(5)=\$235$ Feb.
$26(7)+31(2)-3(3)-2(5)=\$225$ Mar.
$178+235+225=\$638$ total

**How Am I Doing? Sections 5.1-5.4**

**1.** $\frac{\overset{4}{\cancel{16}}}{\cancel{9}}\cdot\frac{-\overset{2}{\cancel{18}}}{\underset{9}{\cancel{36}}}=-\frac{8}{9}$

**2.** $\frac{-\cancel{3}y^3}{\underset{5}{\cancel{20}}}\cdot\frac{-\overset{3}{\cancel{12}}y^2}{\underset{7}{\cancel{21}}}=\frac{3y^5}{35}$

**3.** $25\div\frac{5}{7}=\overset{5}{\cancel{25}}\cdot\frac{7}{\cancel{5}}=35$

**4.** $\frac{3y^4}{20}\div\frac{12y^2}{5}=\frac{\cancel{3}y^{\overset{2}{\cancel{4}}}}{\underset{4}{\cancel{20}}}\cdot\frac{\cancel{5}}{\underset{4}{\cancel{12}}\cancel{y^2}}=\frac{y^2}{16}$

**5.** $63\div\frac{3}{4}=63\cdot\frac{4}{3}=84$ parcels

**6.** $12=2^2\cdot 3,\ 21=3\cdot 7,\ \text{LCM}=2^2\cdot 3\cdot 7=84$

**7.** $7x = 7x,\ 21 = 3\cdot 7,\ 2x^2 = 2x^2$
$\text{LCM} = 2\cdot 3\cdot 7x^2 = 42x^2$

**8.** $\frac{-3}{7}\cdot\frac{3}{3}+\frac{2}{21}=\frac{-9+2}{21}=-\frac{7}{21}=-\frac{1}{3}$

**9.** $\frac{11}{16}\cdot\frac{5}{5}-\frac{5}{20}\cdot\frac{4}{4}=\frac{55-20}{80}=\frac{35}{80}=\frac{7}{16}$

**10.** $\frac{5x}{8}\cdot\frac{7}{7}-\frac{3x}{14}\cdot\frac{4}{4}=\frac{35x-12x}{56}=\frac{23x}{56}$

**11.** (a) $\frac{1}{4}\cdot\frac{5}{5}-\frac{1}{5}\cdot\frac{4}{4}=\frac{5-4}{20}=\frac{1}{20}$

(b) $\frac{1}{4}\cdot\frac{5}{5}+\frac{1}{5}\cdot\frac{4}{4}=\frac{5+4}{20}=\frac{9}{20}$

**12.** $2\frac{2}{5}+4\frac{6}{7}=\frac{12}{5}\cdot\frac{7}{7}+\frac{34}{7}\cdot\frac{5}{5}=\frac{254}{35}=7\frac{9}{35}$

**13.** $6\frac{1}{12}-2\frac{7}{15}=\frac{73}{12}\cdot\frac{5}{5}-\frac{37}{15}\cdot\frac{4}{4}=\frac{217}{60}=3\frac{37}{60}$

**14.** $2\frac{2}{3}\cdot 1\frac{5}{16}=\frac{8}{3}\cdot\frac{21}{16}=\frac{7}{2}=3\frac{1}{2}$

**15.** $10\frac{2}{9}\div 2\frac{1}{3}=\frac{92}{9}\div\frac{7}{3}=\frac{92}{9}\cdot\frac{3}{7}=\frac{92}{21}=4\frac{8}{21}$

## 5.5 Exercises

**1.** Multiply

**3.** $\frac{3}{5}-\frac{1}{3}\div\frac{5}{6}=\frac{3}{5}-\frac{1}{\cancel{3}}\cdot\frac{\overset{2}{\cancel{6}}}{5}=\frac{3}{5}-\frac{2}{5}=\frac{1}{5}$

**5.** $\frac{3}{4}\cdot\frac{5}{5}+\frac{1}{4}\cdot\frac{3}{5}=\frac{15+3}{20}=\frac{18}{20}=\frac{\cancel{2}\cdot 9}{\cancel{2}\cdot 10}=\frac{9}{10}$

**7.** $\frac{5}{7}\cdot\frac{1}{3}\div\frac{2}{7}=\frac{5}{\underset{3}{\cancel{21}}}\cdot\frac{\cancel{7}}{2}=\frac{5}{6}$

**9.** $\left(\frac{1}{2}\right)^2-\frac{2}{6}+\frac{1}{3}=\frac{1}{4}\cdot\frac{3}{3}-\frac{2}{6}\cdot\frac{2}{2}+\frac{1}{3}\cdot\frac{4}{4}$
$=\frac{3-4+4}{12}=\frac{3}{12}=\frac{1}{4}$

**11.** $\frac{5}{6}\cdot\frac{1}{2}+\frac{2}{3}\div\frac{4}{3}=\frac{5}{12}+\frac{2}{3}\cdot\frac{3}{4}=\frac{5}{12}+\frac{1}{2}\cdot\frac{6}{6}$
$=\frac{5+6}{12}=\frac{11}{12}$

**13.** $\frac{\cancel{2}}{9}\cdot\frac{1}{\underset{2}{\cancel{4}}}+\left(\frac{2}{3}\div\frac{6}{7}\right)=\frac{1}{18}+\left(\frac{\cancel{2}}{3}\cdot\frac{7}{\underset{3}{\cancel{6}}}\right)$
$=\frac{1}{18}+\frac{7}{9}\cdot\frac{2}{2}=\frac{1+14}{18}=\frac{\overset{5}{\cancel{15}}}{\underset{6}{\cancel{18}}}=\frac{5}{6}$

**15.** $\frac{3}{4}\cdot\frac{1}{4}+\left(\frac{3}{4}\right)^2=\frac{3}{16}+\frac{9}{16}=\frac{12}{16}=\frac{3}{4}$

**17.** $\left(-\frac{2}{5}\right)\cdot\left(\frac{1}{4}\right)^2=-\frac{2}{5}\cdot\frac{1}{16}=-\frac{1}{40}$

**19.** $\frac{7+(-3)^2}{\frac{8}{9}}=(7+9)\cdot\frac{9}{8}=\overset{2}{\cancel{16}}\cdot\frac{9}{\cancel{8}}=18$

**21.** $\frac{\frac{4}{7}}{2^3+8}=\frac{\frac{4}{7}}{8+8}=\frac{\frac{4}{7}}{16}=\frac{\cancel{4}}{7}\cdot\frac{1}{\underset{4}{\cancel{16}}}=\frac{1}{28}$

**23.** $\frac{2\cdot 3-1}{\frac{5}{8}}=(6-1)\cdot\frac{8}{5}=(5)\cdot\frac{8}{5}=8$

**25.** $\dfrac{\frac{6}{7}}{\frac{9}{14}} = \dfrac{\overset{2}{\cancel{6}}}{\cancel{7}} \cdot \dfrac{\overset{2}{\cancel{14}}}{\underset{3}{\cancel{9}}} = \dfrac{4}{3}$

**27.** $\dfrac{\frac{x^2}{3}}{\frac{x}{6}} = \dfrac{x^{\cancel{2}}}{\cancel{3}} \cdot \dfrac{\overset{2}{\cancel{6}}}{\cancel{x}} = 2x$

**29.** $\dfrac{\frac{x}{4}}{\frac{x^2}{12}} = \dfrac{\cancel{x}}{\cancel{4}} \cdot \dfrac{\overset{3}{\cancel{12}}}{x^{\cancel{2}}} = \dfrac{3}{x}$

**31.** $\dfrac{\frac{1}{2}+\frac{3}{4}}{\frac{4}{5}+\frac{1}{10}} \cdot \dfrac{20}{20} = \dfrac{10+15}{16+2} = \dfrac{25}{18}$

**33.** $\dfrac{\frac{4}{25}-\frac{3}{50}}{\frac{3}{10}+\frac{5}{20}} \cdot \dfrac{100}{100} = \dfrac{16-6}{30+25} = \dfrac{10}{55} = \dfrac{2}{11}$

**35.** $\dfrac{\frac{x}{8}}{\frac{x^2}{16}} = \dfrac{\cancel{x}}{\cancel{8}} \cdot \dfrac{\overset{2}{\cancel{16}}}{x^{\cancel{2}}} = \dfrac{2}{x}$

**37.** $\left(\dfrac{1}{2}\right)^2 + \dfrac{2}{3} \cdot \dfrac{6}{7} = \dfrac{1}{4} \cdot \dfrac{7}{7} + \dfrac{4}{7} \cdot \dfrac{4}{4} = \dfrac{23}{28}$

**39.** $\dfrac{\frac{1}{9}+\frac{2}{3}}{\frac{3}{2}+\frac{1}{3}} \cdot \dfrac{18}{18} = \dfrac{2+12}{27+6} = \dfrac{14}{33}$

**41.** $\dfrac{9+3^2}{\frac{2}{3}} = (9+9) \cdot \dfrac{3}{2} = 9 \cdot 3 = 27$

**43.** $\dfrac{2\frac{3}{4}\text{ cups}}{20\text{ people}} \cdot 36\text{ people} = 4\dfrac{19}{20}\text{ cups}$

**45.** $\dfrac{3\frac{3}{4}\text{ bags}}{2\text{ lawns}} \cdot 12\text{ lawns} = 22\dfrac{1}{2}\text{ bags}$

**47.** $\dfrac{2\frac{1}{4}\text{ acres}}{3\text{ homes}} \cdot 28\text{ homes} = 21\text{ acres}$

**49.** $\dfrac{\frac{25xy^2}{49}}{\frac{15x^2y}{14}} = \dfrac{\overset{5}{\cancel{25}}\,\cancel{x}y^{\cancel{2}}}{\underset{7}{\cancel{49}}} \cdot \dfrac{\overset{2}{\cancel{14}}}{\underset{3}{\cancel{15}}\,x^{\cancel{2}}\cancel{y}} = \dfrac{10y}{21x}$

**51.** $\dfrac{\frac{1}{x}-\frac{1}{y}}{\frac{1}{x}+\frac{1}{y}} \cdot \dfrac{xy}{xy} = \dfrac{y-x}{y+x}$

**53.** $x - \dfrac{2}{5} \div \dfrac{4}{15}\Big|_{x=\frac{7}{2}} = \dfrac{7}{2} \cdot \dfrac{5}{5} - \dfrac{\cancel{2}}{5} \cdot \dfrac{15}{\underset{2}{\cancel{4}}} = \dfrac{35-15}{10}$

$= \dfrac{20}{10} = 2$

**55.** $-\dfrac{3}{8} \cdot \dfrac{16}{21} + x\Big|_{x=-\frac{4}{7}} = -\dfrac{\cancel{3}}{\cancel{8}} \cdot \dfrac{\overset{2}{\cancel{16}}}{\underset{7}{\cancel{21}}} + \left(-\dfrac{4}{7}\right)$

$= -\dfrac{2}{7} - \dfrac{4}{7} = -\dfrac{6}{7}$

## Cumulative Review

**57.** $(-2)(3)(-1)(-5) = (-6)(5) = -30$

**59.** $-25 \div 5 = -5$

### 5.6 Exercises

**1.** $\frac{1}{2}+3\cdot\frac{1}{4}=\frac{2}{4}+\frac{3}{4}=\frac{5}{4}=1\frac{1}{4}$ hr

**3.** $d=rt\Rightarrow r=\frac{d}{t}=\frac{310}{5\frac{1}{3}}=310\cdot\frac{3}{16}=\frac{465}{8}$

$r=58\frac{1}{8}$ mph

**5.** $6\cdot1\frac{1}{4}=6\cdot\frac{5}{4}=\frac{15}{2}=7\frac{1}{2}$ cups water

$6\cdot\frac{1}{8}=\frac{6}{8}=\frac{3}{4}$ tsp salt

$6\cdot\frac{1}{4}=\frac{6}{4}=\frac{3}{2}=1\frac{1}{2}$ cups cereal

**7.** $5\cdot2\frac{3}{4}=5\cdot\frac{11}{4}=\frac{55}{4}=13\frac{3}{4}$ yd

**9.** (a) $15\frac{2}{3}\cdot\frac{2}{2}=15\frac{4}{6}$

$-9\frac{1}{2}\cdot\frac{3}{3}=-9\frac{3}{6}$

$6\frac{1}{6}$ gal

(b) $\frac{24\text{ mi}}{\text{gal}}\cdot6\frac{1}{6}\text{ gal}=24\cdot\frac{37}{6}=148$ mi

**11.** $60-\frac{1}{4}\cdot60-\frac{1}{3}\cdot60=25$ ft

**13.** (a) $40-2\cdot4\frac{3}{4}=30\frac{1}{2},\ 25-2\cdot4\frac{3}{4}=15\frac{1}{2}$

dimensions: $30\frac{1}{2}$ ft by $15\frac{1}{2}$ ft

(b) $\left(2\cdot30\frac{1}{2}+2\cdot15\frac{1}{2}\right)\cdot2\frac{1}{4}=\$207$

**15.** $3\cdot4\cdot3\frac{3}{4}=45$ ft, $\frac{45}{8}=5\frac{5}{8}\rightarrow6$ boards

**17.** (a) $50-2\cdot5\frac{1}{2}=39$ ft long

$30-2\cdot5\frac{1}{2}=19$ ft wide

(b) $\frac{\$2\frac{1}{2}}{\text{ft}}\cdot(2\cdot39+2\cdot19)\text{ ft}=\$290$

**19.** $\frac{\frac{10\text{ ft}}{\text{board}}}{\frac{2\frac{3}{4}\text{ ft}}{\text{picket}}}=\frac{3\frac{7}{11}\text{ picket}}{\text{board}}\rightarrow\frac{\frac{56\text{ picket}}{3\text{ picket}}}{\text{board}}$

$=18\frac{2}{3}$ boards $\rightarrow$ 19 boards

**21.** (a) $\frac{240{,}000\text{ mi}}{4\frac{1}{2}\text{ day}}=53{,}333\frac{1}{3}$ mi/day

(b) $\frac{240{,}000\text{ mi}}{4\frac{1}{2}\text{ day}}\cdot\frac{\text{day}}{24\text{ hr}}=2222\frac{2}{9}$ mph

### Cumulative Review

**23.** $3x=12,\ x=\frac{12}{3}=4$

check: $3(4)\overset{?}{=}12,\ 12=12$

**25.** $x-5=12\Rightarrow x=12+5=17$

check: $17-5\overset{?}{=}12,\ 12=12$

**27.** $2(20)+2(16)=72$ ft

## 5.7 Exercises

**1.** When you multiply a nonzero fraction by its reciprocal, the product is <u>1</u>.

**3.** To solve $\frac{x}{6} = 2$, we <u>multiply</u> by 6 on both sides of the equation.

**5.** $\frac{\boxed{12}x}{12} = -3\boxed{12},\ x = \boxed{-36}$

**7.** $\frac{\boxed{-5}x}{-5} = 4\boxed{-5},\ x = \boxed{-20}$

**9.** $\frac{\boxed{5}2x}{\boxed{2}5} = -8 \cdot \frac{\boxed{5}}{\boxed{2}},\ x = \boxed{-20}$

**11.** $\frac{\boxed{7} \cdot -5x}{\boxed{-5}7} = 10 \cdot \frac{\boxed{7}}{\boxed{-5}},\ x = \boxed{-14}$

**13.** $\frac{y}{15} = 12 \Rightarrow y = 15(12) = 180$

check: $\frac{180}{15} \stackrel{?}{=} 12,\ 12 = 12$

**15.** $\frac{x}{7} = 31 \Rightarrow x = 7(31) = 217$

check: $\frac{217}{7} \stackrel{?}{=} 31,\ 31 = 31$

**17.** $\frac{m}{13} = -30 \Rightarrow m = 13(-30) = -390$

check: $\frac{-390}{13} \stackrel{?}{=} -30,\ -30 = -30$

**19.** $\frac{x}{14} = -6 \Rightarrow x = -6(14) = -84$

check: $\frac{-84}{14} \stackrel{?}{=} -6,\ -6 = -6$

**21.** $-15 = \frac{a}{4} \Rightarrow a = 4(-15) = -60$

check: $-15 \stackrel{?}{=} \frac{-60}{4},\ -15 = -15$

**23.** $-4 = \frac{a}{-20} \Rightarrow a = -4(-20) = 80$

check: $-4 \stackrel{?}{=} \frac{80}{-20},\ -4 = -4$

**25.** $\frac{x}{3^2} = 2 + 6 \div 3 \Rightarrow \frac{x}{9} = 2 + 2 = 4$

$x = 9(4) = 36$

check: $\frac{36}{3^2} \stackrel{?}{=} 2 + 6 \div 3,\ 4 = 4$

**27.** $\frac{x}{-7} = -2 + 3 = 1 \Rightarrow x = -7 \cdot 1 = -7$

check: $\frac{-7}{-7} \stackrel{?}{=} -2 + 3,\ 1 = 1$

**29.** $\frac{y}{2^3} = 2 \cdot 3 + 1 \Rightarrow \frac{y}{8} = 6 + 1 = 7$

$y = 8(7) = 56$

check: $\frac{56}{2^3} \stackrel{?}{=} 2 \cdot 3 + 1,\ 7 = 7$

**31.** $\frac{3}{4}y = 9 \Rightarrow y = \frac{4}{3} \cdot 9 = 12$

check: $\frac{3}{4} \cdot 12 \stackrel{?}{=} 9,\ 9 = 9$

**33.** $\frac{2}{7}x = 12 \Rightarrow x = \frac{7}{2} \cdot 12 = 42$

check: $\frac{2}{7} \cdot 42 \stackrel{?}{=} 12,\ 12 = 12$

**35.** $\frac{1}{2}x = -15 \Rightarrow x = 2(-15) = -30$

check: $\frac{1}{2}(-30) \stackrel{?}{=} -15,\ -15 = -15$

**37.** $\frac{-5}{8}x = 30 \Rightarrow x = \frac{8}{-5} \cdot 30 = -48$

check: $\frac{-5}{8} \cdot (-48) \stackrel{?}{=} 30,\ 30 = 30$

**39.** $3 = \frac{y}{11} \Rightarrow y = 3(11) = 33$

check: $3 \stackrel{?}{=} \frac{33}{11},\ 3 = 3$

**41.** $\frac{x}{3^2} = 4 + 8 \div 2,\ \frac{x}{9} = 8,\ x = 9 \cdot 8 = 72$

check: $\frac{72}{3^2} \stackrel{?}{=} 4 + 8 \div 2,\ 8 = 8$

**43.** $\frac{-3}{4}y = -12 \Rightarrow y = \frac{4}{-3} \cdot (-12) = 16$

check: $\frac{-3}{4} \cdot 16 \stackrel{?}{=} -12,\ -12 = -12$

**45.** $-1 = \frac{m}{30} \Rightarrow m = -1(30) = -30$

check: $-1 \stackrel{?}{=} \frac{-30}{30},\ -1 = -1$

**47.** $\frac{a}{b}x = 6,\ \frac{3}{2} \cdot \frac{a}{b}x = 6 \cdot \frac{3}{2} = 9$

$x = 9$ for $\frac{a}{b} = \frac{2}{3}$

**49.** $\frac{a}{b}x = 9$

$\frac{7}{3} \cdot \frac{a}{b}x = 9 \cdot \frac{7}{3} = 21$

$x = 21$ for $\frac{a}{b} = \frac{3}{7}$

**51.** (a) $4x = 52,\ x = \frac{52}{4} = 13$

check: $4(13) \stackrel{?}{=} 52,\ 52 = 52$

(b) $\frac{x}{4} = 52 \Rightarrow x = 4(52) = 208$

check: $\frac{208}{4} \stackrel{?}{=} 52,\ 52 = 52$

**53.** (a) $4x = -52,\ x = \frac{-52}{4} = -13$

check: $4(-13) \stackrel{?}{=} -52,\ -52 = -52$

(b) $\frac{x}{4} = -52 \Rightarrow x = 4(-52) = -208$

check: $\frac{-208}{4} \stackrel{?}{=} -52,\ -52 = -52$

**55.** (a) $x - 7 = 21 \Rightarrow x = 21 + 7 = 28$

check: $28 - 7 \stackrel{?}{=} 21,\ 21 = 21$

(b) $x + 12 = 33,\ x = 33 - 12 = 21$

check: $21 + 12 \stackrel{?}{=} 33,\ 33 = 33$

(c) $5x = 3,\ x = \frac{3}{5}$, check: $5 \cdot \frac{3}{5} \stackrel{?}{=} 3,\ 3 = 3$

(d) $\frac{x}{5} = 11,\ x = 5(11) = 55$

check: $\frac{55}{5} \stackrel{?}{=} 11,\ 11 = 11$

**57.** (a) $x - 7 = -12,\ x = -12 + 7 = -5$

check: $-5 - 7 \stackrel{?}{=} -12,\ -12 = -12$

(b) $3x = -2,\ x = -\frac{2}{3}$

check: $3 \cdot \frac{-2}{3} \stackrel{?}{=} -2,\ -2 = -2$

(c) $\frac{x}{6} = -9,\ x = 6(-9) = -54$

check: $\frac{-54}{6} \stackrel{?}{=} -9,\ -9 = -9$

(d) $x + 11 = -34,\ x = -34 - 11 = -45$

check: $-45 + 11 \stackrel{?}{=} -34,\ -34 = -34$

### Cumulative Review

**59.** $145 \div 11 = 13$ R2 or $13\frac{2}{11}$

**61.** $29{,}441 = 29{,}400$ to nearest hundred

**63.** $A = LW = \left(4\frac{2}{3}\right)\left(3\frac{3}{8}\right) = \left(\frac{14}{3}\right)\left(\frac{27}{8}\right)$

$A = 15\frac{3}{4}\text{ ft}^2$

## Chapter 5 Review

**1.** LCM – The least Common Multiple is the smallest common multiple.

**2.** LCD – The Least Common Denominator is the LCM of the denominators.

**3.** Complex Fraction – A fraction that contains at least one fraction in the numerator or denominator.

**4.** The reciprocal of $\frac{2}{-9}$ is $\frac{-9}{2} = -\frac{9}{2}$

**5.** The reciprocal of 7 is $\frac{1}{7}$

**6.** The reciprocal of $\frac{-a}{b}$ is $\frac{b}{-a} = -\frac{b}{a}$

**7.** $\frac{5}{21}\cdot\frac{3}{15} = \frac{15}{21\cdot 15} = \frac{1}{21}$

**8.** $\frac{-\cancel{6}}{\underset{5}{\cancel{35}}}\cdot\frac{\overset{2}{\cancel{14}}}{\underset{3}{\cancel{18}}} = -\frac{2}{15}$

**9.** $\frac{\cancel{9}\cancel{x}}{\underset{5}{\cancel{15}}}\cdot\frac{\overset{7}{\cancel{21}}}{\underset{2}{\cancel{18}}x^{\overset{2}{\cancel{3}}}} = \frac{7}{10x^2}$

**10.** $\frac{\overset{4}{\cancel{8}}x^{\overset{2}{\cancel{3}}}}{\underset{5}{\cancel{25}}}\cdot\frac{-\overset{\cancel{9}}{\cancel{45}}}{\underset{\cancel{9}}{\cancel{18}}\cancel{x}} = -\frac{4x^2}{5}$

**11.** $\frac{-5x^4}{\cancel{6}}\cdot\overset{2}{\cancel{12}}x^3 = -10x^{4+3} = -10x^7$

**12.** $\frac{9}{14}\div\frac{45}{12} = \frac{\cancel{9}}{\underset{7}{\cancel{14}}}\cdot\frac{\overset{6}{\cancel{12}}}{\underset{5}{\cancel{45}}} = \frac{6}{35}$

**13.** $\frac{7}{15}\div\frac{-35}{20} = -\frac{\cancel{7}}{\underset{3}{\cancel{15}}}\cdot\frac{\overset{4}{\cancel{20}}}{\underset{5}{\cancel{35}}} = -\frac{4}{15}$

**14.** $\frac{8}{42}\div\frac{-22}{7} = -\frac{\overset{\overset{2}{\cancel{4}}}{\cancel{8}}}{\underset{\underset{3}{\cancel{6}}}{\cancel{42}}}\cdot\frac{\cancel{7}}{\underset{11}{\cancel{22}}} = -\frac{2}{33}$

**15.** $\frac{11x^5}{25}\div\frac{3}{5x^2} = \frac{11x^5}{\underset{5}{\cancel{25}}}\cdot\frac{\cancel{5}x^2}{3} = \frac{11x^7}{15}$

**16.** $\frac{16x^2}{9}\div\frac{24}{6x^4} = \frac{\overset{4}{\cancel{16}}x^2}{9}\cdot\frac{\cancel{6}x^4}{\underset{\cancel{4}}{\cancel{24}}} = \frac{4x^6}{9}$

**17.** $A = \frac{1}{2}bh = \frac{1}{2}\cdot 18(7) = 63\text{ m}^2$

**18.** $A = \frac{1}{2}bh = \frac{1}{2}\cdot 13(20) = 130\text{ in.}^2$

**19.** $\frac{2}{7}\cdot 3500 = \$1000$ withheld

**20.** $\frac{1}{3}\cdot\frac{1}{2} = \frac{1}{6}$ lb in each container

**21.** $7 = 7,\ 14 = 2\cdot 7,\ \text{LCM} = 2\cdot 7 = 14$

**22.** $10 = 2\cdot 5,\ 20 = 2^2\cdot 5,\ \text{LCM} = 2^2\cdot 5 = 20$

**23.** $18 = 2\cdot 3^2,\ 20 = 2^2\cdot 5$

$\text{LCM} = 2^2\cdot 3^2\cdot 5 = 180$

**24.** $42 = 2\cdot 3\cdot 7,\ 12 = 2^2\cdot 3$

$\text{LCM} = 2^2\cdot 3\cdot 7 = 84$

**25.** $4x = 2^2x,\ 8 = 2^3,\ 16x = 2^4x$

$\text{LCM} = 2^4x = 16x$

**26.** $7x = 7x,\ 14x = 2\cdot 7x,\ 20 = 2^2\cdot 5$
$\text{LCM} = 2^2\cdot 5\cdot 7x = 140x$

**27.** $18x = 2\cdot 3^2 x,\ 45x^2 = 3^2\cdot 5x^2$
$\text{LCM} = 2\cdot 3^2\cdot 5x^2 = 90x^2$

**28.** $20x = 2^2\cdot 5x,\ 25x^2 = 5^2 x^2$
$\text{LCM} = 2^2\cdot 5^2 x^2 = 100x^2$

**29.** $\frac{6}{17} - \frac{3}{17} = \frac{6-3}{17} = \frac{3}{17}$

**30.** $\frac{-23}{27} + \frac{-11}{27} = \frac{-23+(-11)}{27} = -\frac{34}{27}$

**31.** $\frac{7}{x} - \frac{5}{x} = \frac{7-5}{x} = \frac{2}{x}$

**32.** $\frac{x}{7} - \frac{5}{7} = \frac{x-5}{7}$

**33.** $\frac{5}{6}\cdot\frac{3}{3} + \frac{4}{9}\cdot\frac{2}{2} = \frac{15+8}{18} = \frac{23}{18}$

**34.** $\frac{15}{32}\cdot\frac{7}{7} - \frac{7}{28}\cdot\frac{8}{8} = \frac{105-56}{224}$
$= \frac{49}{224} = \frac{7}{32}$

**35.** $\frac{-3}{14}\cdot\frac{3}{3} + \frac{7}{21}\cdot\frac{2}{2} = \frac{-9+14}{42} = \frac{5}{42}$

**36.** $\frac{5}{2x}\cdot\frac{3}{3} + \frac{8}{3x}\cdot\frac{2}{3} = \frac{15+16}{6x} = \frac{31}{6x}$

**37.** $\frac{4x}{15}\cdot\frac{3}{3} + \frac{3x}{45} = \frac{12x+3x}{45} = \frac{15x}{45} = \frac{x}{3}$

**38.** $\frac{3x}{14}\cdot\frac{3}{3} - \frac{5x}{42} = \frac{9x-5x}{42} = \frac{4x}{42} = \frac{2x}{21}$

**39.** $10\frac{1}{2} = 10\frac{5}{10}$
$+3\frac{4}{5} = +3\frac{8}{10}$
$13\frac{13}{10} = 14\frac{3}{10}$

**40.** $12\frac{7}{9} \quad 12\frac{7}{9}$
$+6\frac{2}{3} = +6\frac{6}{9}$
$18\frac{13}{9} = 19\frac{4}{9}$

**41.** $11\frac{1}{5} = 11\frac{5}{25} = 10\frac{30}{25}$
$-6\frac{11}{25} \qquad -6\frac{11}{25}$
$4\frac{19}{25}$

**42.** $25 = 24\frac{14}{14}$
$-16\frac{5}{14}$
$8\frac{9}{14}$

**43.** $4\frac{1}{2}\cdot 2\frac{2}{9} = \frac{\cancel{9}}{\cancel{2}}\cdot\frac{\overset{10}{\cancel{20}}}{\cancel{9}} = 10$

**44.** $2\frac{3}{4} \div 1\frac{3}{7} = \frac{11}{4}\cdot\frac{7}{10} = \frac{77}{40}$

**45.** $4\frac{2}{5} \div 8\frac{1}{3} = \frac{22}{5}\cdot\frac{3}{25} = \frac{66}{125}$

**46.** $-12 \div \frac{2}{3} = -12 \cdot \frac{3}{2} = -18$

**47.** $75 - 2 \cdot 20\frac{1}{2} - 2 \cdot 10\frac{1}{4} = 13\frac{1}{2}$ ft

**48.** $\frac{1}{4} \cdot 7\frac{1}{2} = \frac{1}{4} \cdot \frac{15}{2} = \frac{15}{8}$

$= 1\frac{7}{8}$ ft

**49.** $\frac{3}{4} + \frac{1}{2} \cdot \frac{2}{7} = \frac{3}{4} \cdot \frac{7}{7} + \frac{1}{7} \cdot \frac{4}{4}$

$= \frac{21+4}{28} = \frac{25}{28}$

**50.** $\left(\frac{3}{4}\right)^2 + \frac{1}{8} \div \frac{1}{2} = \frac{9}{16} + \frac{1}{8} \cdot \frac{2}{1}$

$= \frac{9}{16} + \frac{1}{4} \cdot \frac{4}{4} = \frac{19+4}{16} = \frac{13}{16}$

**51.** $\frac{(-2)^2 + 12}{\frac{4}{7}} = \frac{16}{1} \cdot \frac{7}{4} = 28$

**52.** $\frac{\frac{2}{3}}{\frac{1}{9}} = \frac{2}{3} \cdot \frac{9}{1} = 6$

**53.** $\frac{\frac{x}{12}}{\frac{x^2}{20}} = \frac{\cancel{x}}{\cancel{12}_{3}} \cdot \frac{\cancel{20}^{5}}{x^{\cancel{2}}} = \frac{5}{3x}$

**54.** $\frac{\frac{1}{2} \cdot \frac{2}{2} + \frac{1}{4}}{\frac{2}{3} \cdot \frac{3}{3} - \frac{1}{9}} = \frac{\frac{3}{4}}{\frac{5}{9}} = \frac{3}{4} \cdot \frac{9}{5} = \frac{27}{20}$

**55.** $\frac{\frac{1}{3} + \frac{2}{5}}{\frac{1}{5} + \frac{1}{10}} \cdot \frac{30}{30} = \frac{10+12}{6+3} = \frac{22}{9}$

**56.** $\frac{3\frac{1}{4} \text{ acres}}{4 \text{ homes}} \cdot 32 \text{ homes} = 26 \text{ acres}$

**57.** (a) $38\frac{1}{4} - 2 \cdot 4\frac{1}{2} = 29\frac{1}{4}$

$25 - 2 \cdot 4\frac{1}{2} = 16$

dimensions: $29\frac{1}{4}$ ft by 16 ft

(b) $\left(2 \cdot 29\frac{1}{4} + 2 \cdot 16\right) \cdot 3\frac{1}{4} = \$294\frac{1}{8}$

**58.** $2\frac{1}{2} - \frac{7}{8} = \frac{5}{2} \cdot \frac{4}{4} - \frac{7}{8}$

$= \frac{20-7}{8} = \frac{13}{8} = 1\frac{5}{8}$ in.

**59.** $\frac{x}{3} = 10,\ x = 3(10) = 30$

check: $\frac{30}{3} \stackrel{?}{=} 10,\ 10 = 10$

**60.** $\frac{y}{2} = -5,\ y = -5(2) = -10$

check: $\frac{-10}{2} \stackrel{?}{=} -5,\ -5 = -5$

**61.** $\frac{x}{-6} = -2,\ x = (-6)(-2) = 12$

check: $\frac{12}{-6} \stackrel{?}{=} -2,\ -2 = -2$

**62.** $\frac{2}{3}y = 16,\ y = \frac{3}{2} \cdot 16 = 24$

check: $\frac{2}{3} \cdot 24 \stackrel{?}{=} 16,\ 16 = 16$

**63.** $\frac{-3}{4}x = 18,\ x = \frac{4}{-3} \cdot 18 = -24$

check: $\frac{-3}{4} \cdot (-24) \stackrel{?}{=} 18,\ 18 = 18$

**64.** $\frac{1}{3}y = 4,\ y = 3(4) = 12$

check: $\frac{1}{3} \cdot 12 \stackrel{?}{=} 4,\ 4 = 4$

**65.** $\frac{x}{2^3} = 2 \cdot 6 + 1 \Rightarrow \frac{x}{8} = 12 + 1 = 13$

$x = 8(13) = 104$

check: $\frac{104}{2^3} \stackrel{?}{=} 2 \cdot 6 + 1,\ 13 = 13$

**66.** $\frac{y}{3^2} = 1 + 8 \div 2 \Rightarrow \frac{y}{9} = 1 + 4 = 5$

$y = 9(5) = 45$

check: $\frac{45}{3^2} \stackrel{?}{=} 1 + 8 \div 2,\ 5 = 5$

**How Am I Doing? Chapter 5 Test**

**1.** $\frac{\cancel{2}}{\cancel{3}} \cdot \frac{\overset{3}{\cancel{9}}}{\underset{5}{\cancel{10}}} = \frac{3}{5}$

**2.** $\frac{-1}{4} \cdot \frac{2}{5} = -\frac{2}{20} = -\frac{1}{10}$

**3.** $\frac{\cancel{7}x^2}{\cancel{8}} \cdot \frac{\overset{2}{\cancel{16}}x}{\underset{2}{\cancel{14}}} = x^3$

**4.** $\frac{-1}{2} \div \frac{1}{4} = -\frac{1}{\cancel{2}} \cdot \frac{\overset{2}{\cancel{4}}}{1} = -2$

**5.** $\frac{2x^8}{5} \div \frac{22x^4}{15} = \frac{2x^8}{5} \cdot \frac{15}{22x^4} = \frac{3x^4}{11}$

**6.** First six multiples of 9: 9,18,27,36,45,54
First six multiples of 6: 6,12,18,24,30,36
18 and 36 are common.

**7.** $14 = 2 \cdot 7,\ 21 = 3 \cdot 7$

$\text{LCM} = 2 \cdot 3 \cdot 7 = 42$

**8.** $5a = 5a,\ 10a^4 = 2 \cdot 5a^4,\ 20a^2 = 2^2 \cdot 5a^2$

$\text{LCM} = 2^2 \cdot 5a^4 = 20a^4$

**9.** $5 = 5,\ 7 = 7,\ 51 = 3 \cdot 17$

$\text{LCM} = 3 \cdot 5 \cdot 7 \cdot 17 = 1785$

**10.** $30 = 2 \cdot 3 \cdot 5,\ 4 = 2^2$

$\text{LCD} = 2^2 \cdot 3 \cdot 5 = 60$

**11.** $\frac{12}{11} + \frac{3}{11} = \frac{12+3}{11} = \frac{15}{11} = 1\frac{4}{11}$

**12.** $\frac{1}{5a} + \frac{3}{a} \cdot \frac{5}{5} = \frac{1+15}{5a} = \frac{16}{5a}$

**13.**
$$\begin{array}{rcr} 4\frac{5}{6} & & 4\frac{5}{6} \\ +3\frac{1}{3} \cdot \frac{2}{2} & = & +3\frac{2}{6} \\ \hline & & 7\frac{7}{6} = 8\frac{1}{6} \end{array}$$

**14.** $\frac{2}{21} \cdot \frac{3}{3} + \frac{5}{9} \cdot \frac{7}{7} = \frac{6+35}{63} = \frac{41}{63}$

**15.** $\frac{7}{12} \cdot \frac{5}{5} - \frac{2}{15} \cdot \frac{4}{4} = \frac{35-8}{60} = \frac{27}{60} = \frac{9}{20}$

**16.** 
$$\begin{array}{rrr} 10\frac{1}{8}\cdot\frac{3}{3} = & 10\frac{3}{24} = & 9\frac{27}{24} \\ -2\frac{2}{3}\cdot\frac{8}{8} = & -2\frac{16}{24} & -2\frac{16}{24} \\ \hline & & 7\frac{11}{24} \end{array}$$

**17.** $\frac{7x}{15}\cdot\frac{4}{4}-\frac{x}{20}\cdot\frac{3}{3}=\frac{28x-3x}{60}=\frac{25x}{60}=\frac{5x}{12}$

**18.** $(-4)\cdot 5\frac{1}{3}=(-4)\cdot\frac{16}{3}=-\frac{64}{3}=-21\frac{1}{3}$

**19.** $\frac{8x}{15}\cdot\frac{25}{12x^3}=\frac{2}{3}\cdot\frac{5}{3x^2}=\frac{10}{9x^2}$

**20.** $1\frac{2}{3}\div 3=\frac{5}{3}\cdot\frac{1}{3}=\frac{5}{9}$

**21.** $2\frac{1}{2}\div\left(-\frac{3}{5}\right)=-\frac{5}{2}\cdot\frac{5}{3}=-\frac{25}{6}=-4\frac{1}{6}$

**22.** $\left(\frac{2}{3}\right)^2+\frac{1}{2}\cdot\frac{1}{4}=\frac{4}{9}\cdot\frac{8}{8}+\frac{1}{8}\cdot\frac{9}{9}=\frac{32+9}{72}=\frac{41}{72}$

**23.** $\frac{\frac{x}{2}}{\frac{x}{4}}=\frac{\cancel{x}}{\cancel{2}}\cdot\frac{\overset{2}{\cancel{4}}}{\cancel{x}}=2$

**24.** $\frac{\frac{1}{3}+\frac{1}{2}}{\frac{5}{9}-\frac{1}{3}}\cdot\frac{18}{18}=\frac{6+9}{10-6}=\frac{15}{4}=3\frac{3}{4}$

**25.** $\frac{x}{-3}=6+2^2,\ x=-3(10)=-30$

**26.** $\frac{4}{7}y=28\Rightarrow y=\frac{7}{4}\cdot 28=49$

**27.** $\frac{x}{4}=-20\Rightarrow x=4(-20)=-80$

**28.** Each 10 ft board will make 2 shelves with $3\text{ ft}=10-2\cdot 3\frac{1}{2}$ left over. Therefore she needs 5 boards to make 10 shelves.

**29.** (a) $98\frac{1}{2}°\text{F}+3°\text{F}-1\frac{1}{2}°\text{F}=100°\text{F}$

(b) $100°\text{F}-98\frac{1}{2}°\text{F}=1\frac{1}{2}°\text{F}$

**30.** (a) length: $35\frac{1}{2}-2(5)=25\frac{1}{2}$ ft

width: $22-2(5)=12$ ft

(b) $\frac{\$1\frac{1}{2}}{\text{ft}}\cdot\left(2\left(25\frac{1}{2}\right)+2(12)\right)\text{ ft}=\$112\frac{1}{2}$

## Cumulative Test for Chapters 1-5

**1.** $3401=3000+400+1$

**2.** $2+x+8=10+x$

**3.** $5x-3x+x+5=(5-3+1)x+5=3x+5$

**4.** $-8r+3-5r-8=(-8-5)r-5=-13r-5$

**5.** $4+x=15\Rightarrow x=15-4=11$

**6.** $3x=15\Rightarrow x=\frac{15}{3}=5$

**7.** $x-8=18\Rightarrow x=18+8=26$

**8.** $(3x^2)(x^3)(x) = 3x^{2+3+1} = 3x^6$

**9.** $5(x+3) = 5x + 5(3) = 5x + 15$

**10.** $8^2 - 10 + 4 = 64 - 10 + 4 = 54 + 4 = 58$

**11.** $-4 + 6 - 9 = 2 - 9 = -7$

**12.** $(-15) \div 3 = -5$

**13.** $(-7)(-1) = 7$

**14.** $(-10x^2)(5x) = -50x^{2+1} = -50x^3$

**15.** $3 - 12 \div (-2) + 4^2 = 3 - 12 \div (-2) + 16$
$= 3 - (-6) + 16$
$= 9 + 16$
$= 25$

**16.** (a) $72 = 2 \cdot 2 \cdot 2 \cdot 3 \cdot 3$ composite
(b) 19 prime

**17.** $4x = 2^2 x,\ 8x^3 = 2^3 x^3,\ 12xy = 2^2 \cdot 3xy$
$\text{LCM} = 2^3 \cdot 3x^3 y = 24x^3 y$

**18.** $24 = 2^3 \cdot 3,\ 30 = 2 \cdot 3 \cdot 5$
$\text{LCM} = 2^3 \cdot 3 \cdot 5 = 120$

**19.** $\dfrac{\frac{2}{x^2}}{\frac{8}{x}} \cdot \dfrac{x^2}{x^2} = \dfrac{2}{8x} = \dfrac{1}{4x}$

**20.** $\dfrac{\frac{3}{4} - \frac{1}{2}}{\frac{1}{6} + \frac{2}{3}} \cdot \dfrac{12}{12} = \dfrac{9-6}{2+8} = \dfrac{3}{10}$

**21.**
$$\begin{array}{rr} 11\frac{1}{5} \cdot \frac{3}{3} = & 11\frac{3}{15} = 10\frac{18}{15} \\ -3\frac{2}{3} \cdot \frac{5}{5} = & -3\frac{10}{15} \\ \hline & 7\frac{8}{15} \end{array}$$

**22.** $\dfrac{1}{6x} + \dfrac{3}{x} \cdot \dfrac{6}{6} = \dfrac{1+18}{6x} = \dfrac{19}{6x}$

**23.** $\dfrac{5x}{12} \cdot \dfrac{5}{5} - \dfrac{x}{10} \cdot \dfrac{6}{6} = \dfrac{25x - 6x}{60} = \dfrac{19x}{60}$

**24.** $\dfrac{\cancel{2}x^4}{\cancel{5}x} \cdot \dfrac{\cancel{10}^{2}\, x^2}{\cancel{4}_{2}} = x^{4+2-1} = x^5$

**25.** $-\dfrac{1}{6} \div \left(-\dfrac{2}{3}\right) = \dfrac{1}{6} \cdot \dfrac{3}{2} = \dfrac{1}{4}$

**26.** $7\dfrac{1}{2} \div \left(-\dfrac{3}{8}\right) = -\dfrac{15}{2} \cdot \dfrac{8}{3} = -20$

**27.** $\dfrac{1}{2} \cdot \dfrac{5}{5} + \dfrac{3}{5} \cdot 2 \cdot \dfrac{2}{2} = \dfrac{5+12}{10} = \dfrac{17}{10} = 1\dfrac{7}{10}$

**28.** $\dfrac{-2}{9}x = 12 \Rightarrow x = \dfrac{9}{-2} \cdot 12 = -54$

**29.** $\dfrac{x}{-6} = 5 \Rightarrow x = (-6)(5) = -30$

**30.** $12\dfrac{1}{2} - 9\dfrac{1}{3} = 12\dfrac{3}{6} - 9\dfrac{2}{6} = 3\dfrac{1}{6}$ gal

**31.** $\dfrac{2}{3} \cdot 5\dfrac{1}{4} = \dfrac{2}{3} \cdot \dfrac{21}{4} = \dfrac{21}{6} = 3\dfrac{1}{2}$ lb

# Chapter 6

## 6.1 Exercises

**1.** To subtract two polynomials we add the opposite of the second polynomial to the first polynomial.

**3.** Because we can replace this negative sign with a $-1$ and then multiply each term by $-1$. Multiplying a term by $-1$ has the effect of changing the sign.

**5.** The terms of $2z^2+4z-2y^4+3$ are $2z^2,\ 4z,\ -2y^4,\ 3$

**7.** The terms of $6x^6-3x^3-3y-7$ are $6x^6,\ -3x^3,\ -3y,\ -7$

**9.** $(7y-3)+(-4y+6)=7y-3-4y+6$
$=3y+3$

**11.** $(2a^2+(-3a)+5)+(4a-3)$
$=2a^2-3a+5+4a-3=2a^2+a+2$

**13.** $(5y^2+2y-5)+(-4y^2+(-8y)+2)$
$=5y^2+2y-5-4y^2-8y+2$
$=y^2-6y-3$

**15.** $-(5x+2y)=-5x-2y$

**17.** $-(-8x+9)=8x-9$

**19.** $-(-3x+6z-5y)=3x-6z+5y$

**21.** $(10x+7)-(3x+5)=10x+7-3x-5$
$=7x+2$

**23.** $(7x+3)-(-4x-6)=7x+3+4x+6$
$=11x+9$

**25.** $(-8a+5)-(4a-3)=-8a+5-4a+3$
$=-12a+8$

**27.** $(3y^2+4y-6)-(4y^2-6y-9)$
$=3y^2+4y-6-4y^2+6y+9$
$=-y^2+10y+3$

**29.** $(2x^2+6x-5)-(6x^2-4x-8)$
$=2x^2+6x-5-6x^2+4x+8$
$=-4x^2+10x+3$

**31.** $(-6z^2+9z-1)-(3z^2+8z-7)$
$=-6z^2+9z-1-3z^2-8z+7$
$=-9z^2+z+6$

**33.** $(2a^2+9a-1)-(-5a^2-8a-4)$
$=2a^2+9a-1+5a^2+8a+4$
$=7a^2+17a+3$

**35.** $(-6x^2-6x-1)-(3x^2+8x+4)$
$=-6x^2-6x-1-3x^2-8x-4$
$=-9x^2-14x-5$

**37.** $3x-2(5x^2-6)-(-2x^2+x-1)$
$=3x-10x^2+12+2x^2-x+1$
$=-8x^2+2x+13$

**39.** $6x-(3x^2+8x+2)+2(-x^2-9)$
$=6x-3x^2-8x-2-2x^2-18$
$=-5x^2-2x-20$

**41.** $(-4x^2+7x+1)-(x^2-7)$
$=-4x^2+7x+1-x^2+7$
$=-5x^2+7x+8$

**43.** $(4x^2-5x-8)+(3x^2-8x-1)$
$=4x^2-5x-8+3x^2-8x-1$
$=7x^2-13x-9$

**45.** $(4x^2+6x-2)-(3x^2+7x+1)+(x^2-1)$
$=4x^2+6x-2-3x^2-7x-1+x^2-1$
$=2x^2-x-4$

**47.** $4(-9x-6)-(3x^2-7x+1)+4x$
$=-36x-24-3x^2+7x-1+4x$
$=-3x^2-25x-25$

**49.** $(-7x^2+3x-4)-(5x^2+7x-1)+(x^2-8)$
$=-7x^2+3x-4-5x^2-7x+1+x^2-8$
$=-11x^2-4x-11$

**51.** $5x-2(4x^2+3x-2)-(2x^2+8x-1)$
$=5x-8x^2-6x+4-2x^2-8x+1$
$=-10x^2-9x+5$

**53.** $7x-2^3(x+3)-(-3)^2(x^2-2x-1)$
$=7x-8x-24-9x^2+18x+9$
$=-9x^2+17x-15$

**55.** $(ax+3)+(2x^2+5x-6)+(8x-2)$
$=2x^2-10x-5$
$ax+3+2x^2+5x-6+8x-2$
$=2x^2-10x-5$
$2x^2+(a+5+8)x-5=2x^2-10x-5$
$\Rightarrow a+13=-10 \Rightarrow a=-23$

**57.**
$(ax^2-bx-7)+(5x^2-2x+3)=9x^2-5x-4$
$(a+5)x^2-(b+2)x-4=9x^2-5x-4$
$a+5=9,\ a=4$
$b+2=5,\ b=3$

## Cumulative Review

**59.** $(2x)(4x)=8x^2$

**61.** $(-4x)(2x^2)=-8x^3$

**63.** $3\frac{1}{2}\cdot\frac{7}{7}-2\frac{2}{7}\cdot\frac{2}{2}=3\frac{7}{14}-2\frac{4}{14}=1\frac{3}{14}$ mi

## 6.2 Exercises

**1.** The sign of the last term, +4, is wrong. It should be −4.

**3.** (a) A <u>monomial</u> is a polynomial with 1 term.
(b) A <u>binomial</u> is a polynomial with 2 terms.
(c) A <u>trinomial</u> is a polynomial with 3 terms.

**5.** First term of product is: $\boxed{-15y^2}$
Second term of product is: $\boxed{+10y}$
Third term of product is: $\boxed{-30}$
Therefore
$-5(3y^2-2y+6)=\boxed{-15y^2+10y-30}$

**7.** We multiply by the first term of the binomial: $x\boxed{(x^2+3x+1)}=\boxed{x^3+3x^2+x}$
Then we multiply the second term of the binomial: $-1\boxed{(x^2+3x+1)}=\boxed{-x^2-3x-1}$
Finally we simplify:
$(x-1)(x^2+3x+1)=\boxed{x^3+2x^2-2x-1}$

**9.** $(x+1)(x-3)$ $F = \boxed{x^2}$

$O = \boxed{-3x}$

$I = \boxed{+1x}$

$L = \boxed{-3}$

$(x+1)(x-3) = \boxed{x^2 - 2x - 3}$

**11.** $8(2x^2 + 3x - 5) = 16x^2 + 24x - 40$

**13.** $-5y(2y^2 - 3y + 6) = -10y^3 + 15y^2 - 30y$

**15.** $-7x(3x - 2y + 4) = -21x^2 + 14xy - 28x$

**17.** $-4y^2(y^6 + 9) = -4y^8 - 36y^2$

**19.** $(x^3 - 5x - 2)(-3x^4) = -3x^7 + 15x^5 + 6x^4$

**21.** $(x-1)(2x^2 - 3x - 2)$
$= 2x^3 - 3x^2 - 2x - 2x^2 + 3x + 2$
$= 2x^3 - 5x^2 + x + 2$

**23.** $(2y+4)(3y^2 + y - 5)$
$= 6y^3 + 2y^2 - 10y + 12y^2 + 4y - 20$
$= 6y^3 + 14y^2 - 6y - 20$

**25.** $(3x-1)(x^2 + 2x + 1)$
$= 3x^3 + 6x^2 + 3x - x^2 - 2x - 1$
$= 3x^3 + 5x^2 + x - 1$

**27.** $(x+6)(x+7) = x^2 + 7x + 6x + 42$
$= x^2 + 13x + 42$

**29.** $(x+3)(x+9) = x^2 + 9x + 3x + 27$
$= x^2 + 12x + 27$

**31.** $(a+6)(a+2) = a^2 + 2a + 6a + 12$
$= a^2 + 8a + 12$

**33.** $(y+4)(y-8) = y^2 - 8y + 4y - 32$
$= y^2 - 4y - 32$

**35.** $(x+2)(x-4) = x^2 - 4x + 2x - 8$
$= x^2 - 2x - 8$

**37.** $(x-2)(x+4) = x^2 + 4x - 2x - 8$
$= x^2 + 2x - 8$

**39.** $(2x+1)(x+2) = 2x^2 + 4x + x + 2$
$= 2x^2 + 5x + 2$

**41.** $(3x-3)(x-1) = 3x^2 - 3x - 3x + 3$
$= 3x^2 - 6x + 3$

**43.** $(2y-1)(y+2) = 2y^2 + 4y - y - 2$
$= 2y^2 + 3y - 2$

**45.** $(2y+1)(y-2) = 2y^2 - 4y + y - 2$
$= 2y^2 - 3y - 2$

**47.** $-5a(2a - 4b - 7) = -10a^2 + 20ab + 35a$

**49.** $-7x^3(x-3) = -7x^4 + 21x^3$

**51.** $(x-2)(x^2 - 3x + 1)$
$= x^3 - 3x^2 + x - 2x^2 + 6x - 2$
$= x^3 - 5x^2 + 7x - 2$

**53.** $(z+5)(z-2) = z^2 - 2z + 5z - 10$
$= z^2 + 3z - 10$

**55.** $(2x+1)(4x^2+2x-8)$
$=8x^3+4x^2-16x+4x^2+2x-8$
$=8x^3+8x^2-14x-8$

**57.** $(y-2)(y+7)=y^2+7y-2y-14$
$=y^2+5y-14$

**59.** (a) $(z+5)(z+1)=z^2+z+5z+5$
$=z^2+6z+5$
(b) $(z-5)(z-1)=z^2-z-5z+5$
$=z^2-6z+5$

**61.** (a) $(x+3)(x-1)=x^2+2x-3$
(b) $(x-3)(x+1)=x^2-2x-3$

**63.** $(x+2)(x-1)+2(3x+3)$
$=x^2+x-2+6x+6=x^2+7x+4$

**65.** $-2x(x^2+3x-1)+(x-2)(x-3)$
$=-2x^3-6x^2+2x+x^2-5x+6$
$=-2x^3-5x^2-3x+6$

**67.** $a(2x-3)=-14x+21$
$2ax-3a=-14x+21 \Rightarrow 2a=-14$
$a=-7$

**Cumulative Review**

**69.** (a) $D=3N$
(b) $21=3N$, $N=7$ nickels

**71.** $-3x^2+5x^2+(-6x^2)=2x^2+(-6x^2)$
$=-4x^2$

**How Am I Doing in 6.1-6.2?**

**1.** $(3y^2+5y-2)+4y-7=3y^2+9y-9$

**2.** $(-7a+5)-(2a-3)=-9a+8$

**3.** $(-2x^2+4x-7)-(5x^2+3x-4)$
$=-7x^2+x-3$

**4.** $2x-3(5x^2+4)+(-3x^2-x+6)$
$=2x-15x^2-12-3x^2-x+6$
$=-18x^2+x-6$

**5.** $(3x^2-6x-8)-(2x^2+7x+5)+(x^2-7)$
$=3x^2-6x-8-2x^2-7x-5+x^2-7$
$=2x^2-13x-20$

**6.** $-8(2a^2-3a+1)=-16a^2+24a-8$

**7.** $-2y(-6y+4x-5)=12y^2-8xy+10y$

**8.** $-4x^2(x^2+6)=-4x^4-24x^2$

**9.** $(y+2)(4y^2+3y-2)$
$=4y^3+3y^2-2y+8y^2+6y-4$
$=4y^3+11y^2+4y-4$

**10.** $(y+2)(y-4)=y^2-4y+2y-8$
$=y^2-2y-8$

**11.** $(x-4)(x-1)=x^2-4x-x+4$
$=x^2-5x+4$

**12.** $(2y+3)(y+4)=2y^2+8y+3y+12$
$=2y^2+11y+12$

**6.3 Exercises**

**1.** $x=$ Mark's age and $3x=$ Jesse's.

**3.** $x=$ Leslie's running record,
$x-20=$ Alice's running record, and
$x-50=$ Shannon's running record.

**5.** Mark drove 500 miles more than Scott.

**7.** The bother's salary $= B$,
Victor's salary $= B - 95$

**9.** Profit the second quarter $= S$,
profit the fourth quarter $= 31,100 + S$

**11.** width $= W$, length $= 2W$

**13.** The height of the tree $= t$,
the height of a pole $= \frac{1}{2}t$

**15.** length $= L$, width $= 2L - 13$

**17.** Late arrivals this year $= x$,
late arrivals last year $= 2x - 420$

**19.** width $= W$, length $= 2W$, height $= 3W$

**21.** $J =$ Janet's height
$J + 2 =$ Wendy's height
$J - 3 =$ Marci's height

**23.** $F =$ length of first side
$F + 4 =$ length of second side
$2F - 10 =$ length of third side

**25.** (a) height of tree $= t$,
height of building $= 4t$
(b) $4t - t$
(c) $4t - t = 3t$

**27.** (a) Number of coins in John's collection $= x$, Number of coins in Carina's collection $= x + 10$, Number of coins in Karla's collection $= x - 6$
(b) $(x+10) + x - (x-6)$
(c) $(x+10) + x - (x-6)$
$= x + 10 + x - x + 6 = x + 16$

**29.** (a) Sam's salary $= x$,
Vu's salary $= x + 125$,
Evan's salary $= x - 80$
(b) $(x+125) + x - (x-80)$
(c) $(x+125) + x - (x-80)$
$= x + 125 + x - x + 80 = x + 205$

**31.** (a) width $= W$, height $= W + 8$,
length $= 2W - 2$
(b) $W + (W+8) + (2W-2)$
(c) $W + (W+8) + (2W-2)$
$= W + W + 8 + 2W - 2 = 4W + 6$

**33.** We can simplify $3x + 1 + 4x$.

**35.** We can simplify and solve
$4 + 2x + 6x = 11$.

**37.** We can solve $4x - 2$. false

**Cumulative Review**

**39.** $2x = 44 \Rightarrow x = \frac{44}{2} = 22$

**41.** $\frac{m}{7} = -5 \Rightarrow m = 7(-5) = -35$

**Putting Your Skills to Work, Problems for Individual Analysis**

**1.** (a)

| Salary Level | Monthly Salary | Formula |
|---|---|---|
| Level 1 | | $s$ |
| Level 2 | | $s + 300$ |
| Level 3 | | $s + 600$ |
| Level 4 | | $s + 900$ |

**1.** (b)

| Salary Level | Monthly Salary | Formula |
|---|---|---|
| Level 1 | $1600 | $s$ |
| Level 2 | $1900 | $s+300$ |
| Level 3 | $2200 | $s+600$ |
| Level 4 | | $s+900$ |

**2.** (a)

| Salary Level | Salary | Formula |
|---|---|---|
| Level 1 | $1500 | $x$ |
| Level 2 | 1700 | $x+200$ |
| Level 3 | 1900 | $x+400$ |
| Level 4 | | $x+500$ |
| Level 5 | | $x+800$ |

(b)

| Salary Level | Salary | Formula |
|---|---|---|
| Level 1 | $1500 | $x$ |
| Level 2 | 1700 | $x+200$ |
| Level 3 | 1900 | $x+400$ |
| Level 4 | | $x+500$ |
| Level 5 | 2300 | $x+800$ |

**3.** Level 5: $x+800$, $1650+800=2450$

**Putting Your Skills to Work, Problems for Group Investigation and Study**

**1.** There are 3 ways to set up the salary schedule.

| Level | Salary | Formula |
|---|---|---|
| Level 1 | 1600 | $s$ |
| Level 2 | 2600 | $s+1000$ |

**1.**

| Level | Salary | Formula |
|---|---|---|
| Level 1 | 1600 | $s$ |
| Level 2 | 1800 | $s+200$ |
| Level 3 | 2000 | $s+400$ |
| Level 4 | 2200 | $s+600$ |
| Level 5 | 2400 | $s+800$ |
| Level 6 | 2600 | $s+1000$ |

| Level | Salary | Formula |
|---|---|---|
| Level 1 | 1600 | $s$ |
| Level 2 | 2100 | $s+500$ |
| Level 3 | 2600 | $s+1000$ |

**2.** (a) Answers may vary.

| | First Arrangement |
|---|---|
| Front | A 1 A 2 |
| Middle | 7 yr 9 yr |
| Back | 12 yr 14 yr empty |

| | Second Arrangement |
|---|---|
| Front | A 1 A 2 |
| Middle | 9 yr 7 yr |
| Back | empty 12 yr 14 yr |

**2.** (a)

| | Third Arrangement |
|---|---|
| Front | A 2 14 yr |
| Middle | A 1 12 yr |
| Back | 7 yr 9 yr empty |

**2.** (b)

| | First Arrangement |
|---|---|
| Front | A 1 A 2 |
| Middle | 7 yr 14 yr |
| Back | 12 yr 9 yr empty |

| | Second Arrangement |
|---|---|
| Front | A 1 14 yr |
| Middle | 9 yr 12 yr |
| Back | empty A 2 7 yr |

### 6.4 Exercises

**1.** The sign in the binomial should be −, not +.

**3.** $8 = 2 \cdot 2 \cdot 2$

$12 = 2 \cdot 2 \cdot 3$

(a) 2, 4 are the common factors

(b) $\text{GCF} = 2 \cdot 2 = 4$

**5.** $4 = 2 \cdot 2$

$12 = 2 \cdot 2 \cdot 3$

$\text{GCF} = 2 \cdot 2 = 4$

**7.** $18 = 2 \cdot 3 \cdot 3$

$21 = 3 \cdot 7$

$\text{GCF} = 3$

**9.** $6 = 2 \cdot 3$

$9 = 3 \cdot 3$

$15 = 3 \cdot 5$

$\text{GCF} = 3$

**11.** $10 = 2 \cdot 5$

$15 = 3 \cdot 5$

$20 = 2 \cdot 2 \cdot 5$

$\text{CGF} = 5$

**13.** (a) $a$ and $c$

(b) $a$ is 3 and $c$ is 1

(c) $a^3c$

**15.** $xy^2 = x \cdot y \cdot y$

$xy^3 = x \cdot y \cdot y \cdot y$

$\text{GCF} = xy^2$

**17.** $a^2b^5 = a \cdot a \cdot b \cdot b \cdot b \cdot b \cdot b$

$a^3b^4 = a \cdot a \cdot a \cdot b \cdot b \cdot b \cdot b$

$\text{GCF} = a^2b^4$

**19.** $a^3bc^2 = a \cdot a \cdot a \cdot b \cdot c \cdot c$

$ac^3 = a \cdot c \cdot c \cdot c$

$\text{GCF} = ac^2$

**21.** $x^3yz^3 = x \cdot x \cdot x \cdot y \cdot z \cdot z \cdot z$

$xy^4 = x \cdot y \cdot y \cdot y \cdot y$

$\text{GCF} = xy$

**23.** $9x + 12 = \boxed{3}(3x + 4)$

**25.** $10xy^2 - 15y = \boxed{5y}(2xy - 3)$

**27.** $2x + 4 = 2(\boxed{x} + \boxed{2})$

**29.** $6x^2 - 12x = 3x(\boxed{2x} - \boxed{4})$

**31.** $18x^3 - 3x^2 - 9x = 3x(\boxed{6x^2} - \boxed{x} - \boxed{3})$

**33.** (a) $2x-10=2(x\boxed{-}5)$
(b) $2x+10=2(x\boxed{+}5)$

**35.** $3a-12=3(a-4)$
check: $3(a-4)=3a-3\cdot 4=3a-12$

**37.** $6y+6=6(y+1)$
check: $6(y+1)=6y+6(1)=6y+6$

**39.** $4a+10b=2(2a+5b)$
check: $2(2a+5b)=4a+10b$

**41.** $3m+15n=3(m+5n)$
check: $3(m+5n)=3m+15n$

**43.** $7x+14y+28=7(x+2y+4)$
check: $7(x+2y+4)=7x+14y+28$

**45.** $8a+18b-6=2(4a+9b-3)$
check: $2(4a+9b-3)=8a+18b-6$

**47.** $2a^2-4a=2a(a-2)$

**49.** $4ab-b^2=b(4a-b)$

**51.** $5x+10xy=5x(1+2y)$

**53.** $7x^2y+14xy=7xy(x+2)$

**55.** $12a^2b-6a^2=6a^2(2b-1)$

**57.** $5x^2+10x^3=5x^2(1+2x)$

**59.** $2x^2y+4xy=2xy(x+2)$

**61.** $4y+2=2(2y+1)$
check: $2(2y+1)=2(2y)+2(1)=4y+2$

**63.** $15a-20=5(3a-4)$
check: $5(3a-4)=15a-5\cdot 4=15a-20$

**65.** $5x-10xy=5x(1-2y)$

**67.** $9xy^3-3xy=3xy(3y^2-1)$

**69.** $6x-3y+12=3(2x-y+4)$

**71.** $4x+8y-4=4(x+2y-1)$

**73.** $2x^3y^3-8x^2y^2=2x^2y^2(xy-4)$

**75.** $6x^2y+2xy+4x=2x(3xy+y+2)$

**77.** (a) $-2(x-5y)=-2x+10y$
(b) $2(-x+5y)=-2x+10y$
(c) The products are the same.

## Cumulative Review

**79.** $3=3,\ 4=2^2,\ 2=2,\ \text{LCD}=2^2\cdot 3=12$

**81.** $2x=2x,\ x=x,\ \text{LCD}=2x$

**83.** $10-2\cdot 2\frac{2}{3}-3\frac{1}{4}$
$=10-2\cdot\frac{8}{3}-\frac{13}{4}$
$=10\cdot\frac{12}{12}-\frac{16}{3}\cdot\frac{4}{4}-\frac{13}{4}\cdot\frac{3}{3}$
$=\frac{120-64-39}{12}=\frac{17}{12}=1\frac{5}{12}$ in.

## Chapter 6 Review Problems

**1.** The terms in $2x^2+5x-3z^3+4$ are $2x^2,\ +5x,\ -3z^3,\ +4$

**2.** The terms in $8a^5 - 7b^3 - 5b - 4$ are $8a^5,\ -7b^3,\ -5b,\ -4$

**3.** $-(3a-4) = -3a+4$

**4.** $-(-3y+2z-3) = 3y-2z+3$

**5.** $(-3x+9)+(5x-2) = -3x+9+5x-2$
$= 2x+7$

**6.** $(4x+8)-(8x-2) = 4x+8-8x+2$
$= -4x+10$

**7.** $(9a^2-3a+5)-(-4a^2-6a-1)$
$= 9a^2-3a+5+4a^2+6a+1$
$= 13a^2+3a+6$

**8.** $(-4x^2-3)-(3x^2+7x+1)+(-x^2-4)$
$= -4x^2-3-3x^2-7x-1-x^2-4$
$= -8x^2-7x-8$

**9.** $2(-2x^2+2)-(3x^2+5x-6)$
$= -4x^2+4-3x^2-5x+6$
$= -7x^2-5x+10$

**10.** $4x-(x^2+2x)+3(3x^2-6x+4)$
$= 4x-x^2-2x+9x^2-18x+12$
$= 8x^2-16x+12$

**11.** $(-4)(6x^2-8x+5) = -24x^2+32x-20$

**12.** $(-2y)(y-6) = -2y^2+12y$

**13.** $(3x)(9x-3y+2) = 27x^2-9xy+6x$

**14.** $(-5n)(-4n-9m-7)$
$= 20n^2+45mn+35n$

**15.** $(4x^2)(x^4-4) = 4x^6-16x^2$

**16.** $(x^4)(x^5-2x-3) = x^9-2x^5-3x^4$

**17.** $(z-4)(5z) = 5z^2-20z$

**18.** $(y+10)(-6y) = -6y^2-60y$

**19.** $(x^3-6x)(4x^2) = 4x^5-24x^3$

**20.** $(x-2)(2x^2+3x-1)$
$= 2x^3+3x^2-x-4x^2-6x+2$
$= 2x^3-x^2-7x+2$

**21.** $(y+5)(3y^2-2y+3)$
$= 3y^3-2y^2+3y+15y^2-10y+15$
$= 3y^3+13y^2-7y+15$

**22.** $(y-1)(-3y^2+4y+5)$
$= -3y^3+4y^2+5y+3y^2-4y-5$
$= -3y^3+7y^2+y-5$

**23.** $(2x+3)(x^2+3x-1)$
$= 2x^3+6x^2-2x+3x^2+9x-3$
$= 2x^3+9x^2+7x-3$

**24.** $(x+2)(x+4) = x^2+4x+2x+8$
$= x^2+6x+8$

**25.** $(y+4)(y-7) = y^2-7y+4y-28$
$= y^2-3y-28$

**26.** $(x-2)(3x+4) = 3x^2+4x-6x-8$
$= 3x^2-2x-8$

**27.** $(x-3)(5x-6) = 5x^2 - 6x - 15x + 18$
$= 5x^2 - 21x + 18$

**28.** Profit for first quarter $= x$,
profit for third quarter $= 22{,}300 + x$

**29.** length $= L$, width $= L - 22$

**30.** Measure of $\angle b = x$,
measure of $\angle a = x + 30$,
measure of $\angle c = 2x$

**31.** Senior citizen tickets sold $= x$,
adult tickets sold $= 3x$,
children's tickets sold $= x + 40$

**32.** (a) Erin's salary $= x$,
Phoebe's salary $= x + 145$,
Kelly's salary $= x - 60$
(b) $x + (x + 145) - (x - 60)$
(c) $x + (x + 145) - (x - 60)$
$= x + x + 145 - x + 60 = x + 205$

**33.** (a) Number of coins in Jessica's collection $= x$,
Number of coins in Natasha's collection $= x + 20$,
Number of coins in Le Mar's collection $= x - 15$
(b) $x + (x + 20) - (x - 15)$
(c) $x + (x + 20) - (x - 15)$
$= x + x + 20 - x + 15 = x + 35$

**34.** (a) first angle $= x$, second angle $= 2x$,
third angle $= x + 10$
(b) $x + 2x + (x + 10)$
(c) $x + 2x + (x + 10) = 3x + x + 10$
$= 4x + 10$

**35.** (a) width $= W$, length $= W + 7$,
height $= 3W - 4$
(b) $W + (W + 7) + (3W - 4)$
(c) $W + (W + 7) + (3W - 4)$
$= W + W + 7 + 3W - 4 = 5W + 3$

**36.**
$7 = 7$
$21 = 3 \cdot 7$
GCF $= 7$

**37.**
$6 = 2 \cdot 3$
$21 = 3 \cdot 7$
GCF $= 3$

**38.**
$25 = 5 \cdot 5$
$40 = 2 \cdot 2 \cdot 2 \cdot 5$
GCF $= 5$

**39.**
$18 = 2 \cdot 3 \cdot 3$
$30 = 2 \cdot 3 \cdot 5$
GCF $= 2 \cdot 3 = 6$

**40.**
$8 = 2 \cdot 2 \cdot 2$
$14 = 2 \cdot 7$
$18 = 2 \cdot 3 \cdot 3$
GCF $= 2$

**41.**
$12 = 2 \cdot 2 \cdot 3$
$16 = 2 \cdot 2 \cdot 2 \cdot 2$
$20 = 2 \cdot 2 \cdot 5$
GCF $= 2 \cdot 2 = 4$

**42.**
$a^2bc = a \cdot a \cdot b \cdot c$
$ab^3 = a \cdot b \cdot b \cdot b$
GCF $= ab$

**43.** $xy^3z = x \cdot \quad y \cdot y \cdot y \cdot z$
$x^2y = x \cdot x \cdot y$
$\text{GCF} = xy$

**44.** $2x - 14 = 2(x - 7)$

**45.** $5x + 15 = 5(x + 3)$

**46.** $3a + 9b = 3(a + 3b)$

**47.** $4y + 8z = 4(y + 2z)$

**48.** $6xy^2 - 12xy = 6xy(y - 2)$

**49.** $8a^2b - 16ab = 8ab(a - 2)$

**50.** $10x^3y + 5x^2y = 5x^2y(2x + 1)$

**51.** $4y^3 - 6y^2 + 2y = 2y(2y^2 - 3y + 1)$

**52.** $3a - 6b + 12 = 3(a - 2b + 4)$

**53.** $2x + 4y - 10 = 2(x + 2y - 5)$

## How Am I Doing in Chapter 6?

**1.** The terms in $x^2y - 2x^2 + 3y - 5x$ are
$x^2y, \ -2x^2, \ +3y, \ -5x$

**2.** $-(4x - 2y - 6) = -4x + 2y + 6$

**3.** $(-5x + 3) + (-2x + 4) = -5x + 3 - 2x + 4$
$= -7x + 7$

**4.** $(4y + 5) - (2y - 3) = 4y + 5 - 2y + 3$
$= 2y + 8$

**5.** $(-7p - 2) - (3p + 4) = -7p - 2 - 3p - 4$
$= -10p - 6$

**6.** $(4x^2 + 8x - 3) + (9x^2 - 10x + 1)$
$= 4x^2 + 8x - 3 + 9x^2 - 10x + 1$
$= 13x^2 - 2x - 2$

**7.** $(-6m^2 - 3m - 8) - (6m^2 + 3m - 4)$
$= -6m^2 - 3m - 8 - 6m^2 - 3m + 4$
$= -12m^2 - 6m - 4$

**8.** $(x^2 - x + 7) + (-2x^2 + 4x + 6) - (x^2 + 8)$
$= x^2 - x + 7 - 2x^2 + 4x + 6 - x^2 - 8$
$= -2x^2 + 3x + 5$

**9.** $3x - 2(7x^2 + 2x - 1) - (3x^2 + 8x - 2)$
$= 3x - 14x^2 - 4x + 2 - 3x^2 - 8x + 2$
$= -17x^2 - 9x + 4$

**10.** $(-7b)(2b - 4) = -14b^2 + 28b$

**11.** $3(6x^2 - x + 1) = 18x^2 - 3x + 3$

**12.** $(-2x^3)(4x^2 - 3) = -8x^5 + 6x^3$

**13.** $(x + 5)(x + 9) = x^2 + 9x + 5x + 45$
$= x^2 + 14x + 45$

**14.** $(x + 3)(x - 2) = x^2 - 2x + 3x - 6$
$= x^2 + x - 6$

**15.** $(3x^3 - 1)(-4x^4) = -12x^7 + 4x^4$

**16.** $(y - 3)(4y^2 + 2y - 6)$
$= 4y^3 + 2y^2 - 6y - 12y^2 - 6y + 18$
$= 4y^3 - 10y^2 - 12y + 18$

**17.** length $= L$, width $= L - 3$

**18.** length of first side $= f$,
length of second side $= f + 6$,
length of third side $= 2f - 2$

**19.** (a)Number of votes for Lena $= x$,
Number of votes for Jason $= x - 3000$
Number of votes for Nhan $= x + 5100$
(b) $(x + 5100) + x - (x - 3000)$
(c) $(x + 5100) + x - (x - 3000)$
$= x + 5100 + x - x + 3000$
$= x + 8100$

**20.** $9 = 3 \cdot 3$
$21 = 3 \cdot \ 7$
$\text{GCF} = 3$

**21.** $8 = 2 \cdot 2 \cdot 2$
$16 = 2 \cdot 2 \cdot 2 \cdot 2$
$20 = 2 \cdot 2 \quad 5$
$\text{GCF} = 2 \cdot 2 = 4$

**22.** $x^2yz = x \cdot x \cdot \ y \cdot z$
$x^3z = x \cdot x \cdot x \cdot \ z$
$\text{GCF} = x \cdot x \cdot z = x^2z$

**23.** $3x + 12 = 3(x + 4)$

**24.** $7x^2 - 14x + 21 = 7(x^2 - 2x + 3)$

**25.** $2x^2y - 6xy^2 = 2xy(x - 3y)$

## Cumulative Test for Chapters 1-6

**1.** 35

**2.** Saturday

**3.** $65 - 20 = 45$

**4.** (a) factors $= 7,\ x$
product $= -21$
(b) factors $= x,\ y$
product $= z$

**5.** $5 - 12(9 - 10)$
$= 5 - 12(-1)$
$= 5 - (-12)$
$= 17$

**6.** $7 - 24 \div 6(-2)^2 - 3$
$= 7 - 24 \div 6(4) - 3$
$= 7 - 4(4) - 3$
$= 7 - 16 - 3$
$= -9 - 3 = -12$

**7.** $-6 - (-1) = -6 + 1 = -5$

**8.** $-9 - 7 + (-10) = -16 + (-10) = -26$

**9.** $9(0)(1)(-3)(-2) = 0$

**10.** $0 > -72$

**11.** $-12 < -1.2$

**12.** $|-6| = -(-6) = 6,\ |3| = 3,\ 6 > 3 \Rightarrow |-6| > |3|$

**13.** $|8| = 8,\ |-8| = -(-8) = 8,\ 8 = 8 \Rightarrow |8| = |-8|$

**14.** $(3x - 2) + (-5x + 4) = 3x - 2 - 5x + 4$
$= -2x + 2$

**15.** $(-8y + 6) - (6y - 7) = -8y + 6 - 6y + 7$
$= -14y + 13$

**16.** $(-8x^2 + 3x - 6) - (x^2 + 5x)$
$= -8x^2 + 3x - 6 - x^2 - 5x = -9x^2 - 2x - 6$

**17.** $4y-3(8y^2+2y-6)-(3y^2+2y-1)$
$=4y-24y^2-6y+18-3y^2-2y+1$
$=-27y^2-4y+19$

**18.** $(-2a)(3a-4)=-6a^2+8a$

**19.** $(-6)(4x^2-x+9)=-24x^2+6x-54$

**20.** $(4y^2+8)(-2y^2)=-8y^4-16y^2$

**21.** $(x+1)(3x^2-2x+6)$
$=3x^3-2x^2+6x+3x^2-2x+6$
$=3x^3+x^2+4x+6$

**22.** $(x+2)(x+6)=x^2+6x+2x+12$
$=x^2+8x+12$

**23.** $(2x+7)(x-3)=2x^2-6x+7x-21$
$=2x^2+x-21$

**24.** height of tree $=t$,
height of building $=2t+4$

**25.** (a) width $=W$, length $=W+6$,
height $=3W$
(b) $3W+W-(W+6)$
(c) $3W+W-(W+6)$
$=4W-W-6$
$=3W-6$

**26.** $3x+12=3(x+4)$

**27.** $9a+18b+9=9(a+2b+1)$

**28.** $5y^2-10y=5y(2y-2)$

**29.** $12x^2y+6x^2=6x^2(2xy+1)$

# Chapter 7

## 7.1 Exercises

**1.** To solve the equation $y+8=-17$ we add the opposite of $+8$, which is $\underline{-8}$, to both sides of the equation.

**3.** To solve the equation $x-15=82$, we add the number $\underline{15}$ to both sides.

**5.** Before we solve the equation $7x+2-6x-9=2-8$, we must first simplify each side of the equation.

**7.** To solve the equation $-9y=-36$ we divide by $-9$ on both sides of the equation.

**9.** To solve the equation $\frac{y}{-9}=2$, we multiply by $-9$ on both sides of the equation.

**11.** We add $-3$ on both sides of the equation when we solve $14=x+3$.

**13.** $x-6=-19 \Rightarrow x=-19+6=-13$

check: $-13-6\stackrel{?}{=}-19,\ -19=-19$

**15.** $a+9=-1 \Rightarrow a=-1-9=-10$

check: $-10+9\stackrel{?}{=}-1,\ -1=-1$

**17.** $-3-5=x+5-5,\ x=-8$

check: $-3\stackrel{?}{=}-8+5,\ -3=-3$

**19.** $6-9+5=y-5+5,\ y=2$

check: $6-9\stackrel{?}{=}2-5,\ -3=-3$

**21.** $7-12=x-6+2^2,\ x-6+4=-5$

$x-2+2=-5+2,\ x=-3$

check: $7-12\stackrel{?}{=}-3-6+2^2,\ -5=-5$

**23.** $-8+4^2=a+6-4,\ -8+16=a+2$

$a+2-2=8-2,\ a=6$

check: $-8+4^2\stackrel{?}{=}6+6-4,\ 8=8$

**25.** $12x-1-11x-1=-6 \Rightarrow x-2=-6$

$x=-4$

check: $12(-4)-1-11(-4)-1\stackrel{?}{=}-6$

$-6=-6$

**27.** $12-16=4y+6-3y \Rightarrow -4=y+6$

$y=-4-6 \Rightarrow y=-10$

check: $12-16\stackrel{?}{=}4(-10)+6-3(-10)$

$-4=-4$

**29.** $6x-6-5x+1=-2+4 \Rightarrow x-5=2$

$x=2+5=7$

check: $6(7)-6-5(7)+1\stackrel{?}{=}-2+4$

$2=2$

**31.** $-16=\frac{x}{3} \Rightarrow x=-16(3)=-48$

check: $-16\stackrel{?}{=}\frac{-48}{3},\ -16=-16$

**33.** $14=\frac{y}{-2} \Rightarrow y=14(-2)=-28$

check: $14\stackrel{?}{=}\frac{-28}{-2},\ 14=14$

**35.** $\frac{a}{-4}=-6+1=-5 \Rightarrow a=-5(-4)=20$

check: $\frac{20}{-4}\stackrel{?}{=}-6+1,\ -5=-5$

**37.** $\frac{x}{4}=-8+6=-2 \Rightarrow x=4(-2)=-8$

check: $\frac{-8}{4}\stackrel{?}{=}-8+6,\ -2=-2$

**39.** $\frac{y}{-2}=4+2^2=4+4=8$

$y=-2(8)=-16$

check: $\frac{-16}{-2}\stackrel{?}{=}4+2^2,\ 8=8$

**41.** $\frac{5}{3}\cdot\frac{3}{5}x=9\cdot\frac{5}{3},\ x=15$

check: $\frac{3}{5}\cdot 15\stackrel{?}{=}9,\ 9=9$

**43.** $\left(-\frac{3}{2}\right)\left(-\frac{2}{3}a\right)=10\left(-\frac{3}{2}\right),\ a=-15$

check: $-\frac{2}{3}(-15)\stackrel{?}{=}10,\ 10=10$

**45.** $\frac{5}{6}y=4^2+4=16+4=20,\ y=20\cdot\frac{6}{5}=24$

check: $\frac{5}{6}\cdot 24\stackrel{?}{=}4^2+4,\ 20=20$

**47.** $3(2x)=18 \Rightarrow 6x=18 \Rightarrow x=3$

check: $3(2(3))\stackrel{?}{=}18,\ 18=18$

**49.** $-6=2(-5x) \Rightarrow -6=-10x$

$x=\frac{-6}{-10}=\frac{3}{5}$

check: $-6\stackrel{?}{=}2\left(-5\left(\frac{3}{5}\right)\right),\ -6=-6$

**51.** $8x+4(4x)=48 \Rightarrow 8x+16x=48$

$24x=48 \Rightarrow x=2$

check: $8(2)+4(4(2))\stackrel{?}{=}48,\ 48=48$

**53.** $6(-2x)-3x=-30 \Rightarrow -12x-3x=-30$

$-15x=-30 \Rightarrow x=2$

check: $6(-2(2))-3(2)\stackrel{?}{=}-30$

$-30=-30$

**55.** $\frac{-10}{2}=3(4x)+2x \Rightarrow 12x+2x=-5$

$14x=-5 \Rightarrow x=\frac{-5}{14}$

check: $\frac{-10}{2}\stackrel{?}{=}3\left(4\cdot\frac{-5}{14}\right)+2\cdot\frac{-5}{14}$

$-5=-5$

**57.** $\frac{21}{3}=5x+3(-4x) \Rightarrow 7=5x-12x$

$-7x=7 \Rightarrow x=-1$

check: $\frac{21}{3}\stackrel{?}{=}5(-1)+3(-4(-1)),\ 7=7$

**59.** $-x=9 \Rightarrow x=-9$

check: $-(-9)\stackrel{?}{=}9,\ 9=9$

**61.** $-x=-3 \Rightarrow x=-(-3)=3$

check: $-(3)\stackrel{?}{=}-3,\ -3=-3$

**63.** (a) $2x=-12 \Rightarrow x=\frac{-12}{2}=-6$

(b) $\frac{x}{2}=-12 \Rightarrow x=2(-12)=-24$

(c) $x-2=-12 \Rightarrow x=-12+2=-10$

(d) $x+2=-12 \Rightarrow x=-12-2=-14$

**65.** (a) $y-10=9 \Rightarrow y=9+10=19$

(b) $y+10=9 \Rightarrow y=9-10=-1$

(c) $\frac{y}{-10}=9 \Rightarrow y=-10(9)=-90$

(d) $-10y=9 \Rightarrow y=\frac{9}{-10}$

**67.** $-16 = a + 4 \Rightarrow a = -16 - 4 = -20$

check: $-16 \stackrel{?}{=} -20 + 4,\ -16 = -16$

**69.** $5x + 2 - 4x = 9 \Rightarrow x = 9 - 2 = 7$

check: $5(7) + 2 - 4(7) \stackrel{?}{=} 9,\ 9 = 9$

**71.** $\frac{x}{7} = -2 + 3^3 = -2 + 9 = 7 \Rightarrow x = 7 \cdot 7 = 49$

check: $\frac{49}{7} \stackrel{?}{=} -2 + 3^2,\ 7 = 7$

**73.** $-x(-1) = 12(-1),\ x = -12$

check: $-(-12) \stackrel{?}{=} 12,\ 12 = 12$

**75.** $\frac{20}{5} = 3x + 2(-6x),\ 4 = 3x - 12x$

$-9x = 4,\ x = -\frac{4}{9}$

check: $\frac{20}{5} \stackrel{?}{=} 3\left(-\frac{4}{9}\right) + 2\left(-6\left(-\frac{4}{9}\right)\right)$

$4 = 4$

**77.** $\frac{6}{7}x = 3^2 + 3 = 9 + 3 = 12, x = \frac{7}{6} \cdot 12 = 14$

check: $\frac{6}{7} \cdot 14 \stackrel{?}{=} 3^2 + 3,\ 12 = 12$

**79.** (a) $-2x = 8 \Rightarrow x = \frac{8}{-2} = -4$

(b) $-2x \cdot \frac{1}{-2} = 8 \cdot \frac{1}{-2} \Rightarrow x = \frac{8}{-2} = -4$

(c) The answers are the same.

(d) Division is defined in terms of multiplication, therefore dividing by $-2$ is equivalent to multiplying by the reciprocal of $-2$, which is $\frac{1}{-2}$.

**81.** $\angle x + 30° = 90° \Rightarrow \angle x = 90° - 30° = 60°$

**83.** $3x + 2x = 5x = 90°$

$x = 18°,\ 2x = 36°,\ 3x = 54°$

## Cumulative Review Problems

**85.** $66 = 2 \cdot 3 \cdot 11$ **87.** $210 = 2 \cdot 3 \cdot 5 \cdot 7$

**89.** $94{,}500{,}000 - 91{,}400{,}000$
$= 3{,}100{,}000$ mi

## 7.2 Exercises

**1.** If 1 is added to double a number, the result is 5. What is the number?

**3.**

$$\begin{array}{rl} & 3x + 6 = 5x + 9 \\ + & \boxed{-3x} \quad \boxed{-3x} \\ \hline & 0 + 6 = 2x + 9 \\ + & \boxed{-9} \quad \boxed{-9} \\ \hline & -3 = 2x + 0 \end{array}$$

$$\frac{-3}{\boxed{2}} = \frac{2x}{\boxed{2}}$$

$$\frac{-3}{2} = x$$

**5.**

$$\begin{array}{rl} & 6 - 2x = 5 - 9x \\ + & \boxed{+9x} \quad \boxed{+9x} \\ \hline & 6 + 7x = 5 + 0 \\ & \boxed{-6} \quad \boxed{-6} \\ \hline & 0 + 7x = -1 \end{array}$$

$$\frac{7x}{\boxed{7}} = \frac{-1}{\boxed{7}}$$

$$x = -\frac{1}{7}$$

**7.** $-4x+2+3x=13$

$$\begin{array}{rl} \boxed{-1}x+2 & =13 \\ + \quad \boxed{-2} & \boxed{-2} \\ \hline \dfrac{-1x}{\boxed{-1}}+0 & =\dfrac{11}{\boxed{-1}} \\ x & =-11 \end{array}$$

**9.** $3x+9=27 \Rightarrow 3x=18 \Rightarrow x=6$

check: $3(6)+9\stackrel{?}{=}27,\ 27=27$

**11.** $5x-10=25 \Rightarrow 5x=35 \Rightarrow x=7$

check: $5(7)-10\stackrel{?}{=}25,\ 25=25$

**13.** $18=4x+10 \Rightarrow 4x=8 \Rightarrow x=2$

check: $18\stackrel{?}{=}4(2)+10,\ 18=18$

**15.** $5x-1=16 \Rightarrow 5x=17 \Rightarrow x=\frac{17}{5}=3\frac{2}{5}$

check: $5\left(\frac{17}{5}\right)-1\stackrel{?}{=}16,\ 16=16$

**17.** $-4y+7=63 \Rightarrow -4y=56 \Rightarrow y=-14$

check: $-4(-14)+7\stackrel{?}{=}63,\ 63=63$

**19.** $-6m-10=88 \Rightarrow -6m=98 \Rightarrow m=-\frac{49}{3}$

check: $-6\left(-\frac{49}{3}\right)-10\stackrel{?}{=}88,\ 88=88$

**21.** $40=-2x-10 \Rightarrow 50=-2x \Rightarrow x=-25$

check: $40\stackrel{?}{=}-2(-25)-10,\ 40=40$

**23.** $-1=-5y-7 \Rightarrow -5y=6 \Rightarrow y=-\frac{6}{5}$

check: $-5\left(-\frac{6}{5}\right)-7\stackrel{?}{=}-1,\ -1=-1$

**25.** $6=4-2x \Rightarrow 2=-2x \Rightarrow x=-1$

check: $6\stackrel{?}{=}4-2(-1),\ 6=6$

**27.** $-3=6-3y \Rightarrow -9=-3y \Rightarrow y=3$

check: $-3\stackrel{?}{=}6-3(3),\ -3=-3$

**29.** $8y+6-2y=18 \Rightarrow 6y=12 \Rightarrow y=2$

check: $8(2)+6-2(2)\stackrel{?}{=}18,\ 18=18$

**31.** $9x-2+2x=6 \Rightarrow 11x=8 \Rightarrow x=\frac{8}{11}$

check: $9\left(\frac{8}{11}\right)-2+2\left(\frac{8}{11}\right)\stackrel{?}{=}6,\ 6=6$

**33.** $7x+8-8x=11 \Rightarrow -x=3 \Rightarrow x=-3$

check: $7(-3)+8-8(-3)\stackrel{?}{=}11,\ 11=11$

**35.** $15x=9x+30 \Rightarrow 6x=30 \Rightarrow x=5$

check: $15(5)\stackrel{?}{=}9(5)+30,\ 45=45$

**37.** $2x=-12x+5 \Rightarrow 14x=5 \Rightarrow x=\frac{5}{14}$

check: $2\left(\frac{5}{14}\right)\stackrel{?}{=}-12\left(\frac{5}{14}\right)+5,\ \frac{5}{7}=\frac{5}{7}$

**39.** $8x+2=5x-4 \Rightarrow 3x=-6 \Rightarrow x=-2$

check: $8(-2)+2\stackrel{?}{=}5(-2)-4,\ -14=-14$

**41.** $11x+20=12x+2 \Rightarrow -x=-18 \Rightarrow x=18$

check: $11(18)+20\stackrel{?}{=}12(18)+2,\ 218=218$

**43.** $-2+y+8=3y+12 \Rightarrow y+6=3y+12$

$-2y=6 \Rightarrow y=-3$

**45.** $13y+9-2y=6y-8 \Rightarrow 11y+9=6y-8$

$5y=-17 \Rightarrow y=-\frac{17}{5}$

**47.** $-9-3x+8=-6x+2-3x$

$-3x-1=-9x+2 \Rightarrow 6x=3 \Rightarrow x=\frac{3}{6}=\frac{1}{2}$

**49.** $-2y+6=12 \Rightarrow -2y=6 \Rightarrow y=-3$

check: $-2(-3)+6\overset{?}{=}12,\ 12=12$

**51.** $3x-1=14 \Rightarrow 3x=15 \Rightarrow x=5$

check: $3(5)-1\overset{?}{=}14,\ 14=14$

**53.** $5x+5-2x=15 \Rightarrow 3x=10 \Rightarrow x=\frac{10}{3}$

check: $5\left(\frac{10}{3}\right)+5-2\left(\frac{10}{3}\right)\overset{?}{=}15,\ 15=15$

**55.** $13x=8x+20 \Rightarrow 5x=20 \Rightarrow x=4$

check: $13(4)\overset{?}{=}8(4)+20,\ 52=52$

**57.** $4x-24=6x-8 \Rightarrow 2x=-16 \Rightarrow x=-8$

check: $4(-8)-24\overset{?}{=}6(-8)-8,\ -56=-56$

**59.** $2x-6+2(4x)+13=-2-5x+(3^2-3)$

$10x+7=-5x+4 \Rightarrow 15x=-3 \Rightarrow x=-\frac{1}{5}$

check: $2\left(-\frac{1}{5}\right)-6+2\left(4\left(-\frac{1}{5}\right)\right)+13$

$\overset{?}{=}-2-5\left(-\frac{1}{5}\right)+(3^2-3),\ 5=5$

**Cumulative Review Problems**

**61.** $-3(x-4)=-3x+12$

**63.** $-5(-1+x)=5-5x$

**65.** $\frac{4}{5}\cdot\frac{3}{3}=\frac{12}{15},\ \frac{2}{3}\cdot\frac{5}{5}=\frac{10}{15} \Rightarrow \frac{4}{5}>\frac{2}{3}$

Israel was overcharged

### 7.3 Exercises

**1.**

$-5(3x+2)+2x=16$

$-15x-10+2x=16$ Use distributive property to remove parentheses

$-13x-10=16$ Combine like terms

$-13x=26$ Add $+10$ to both sides

$x=-2$ Divide both sides by $-13$

**3.** $-5(2x+1)=25 \Rightarrow -10x-5=25$

$-10x=30 \Rightarrow x=-3$

check: $-5(2(-3)+1)\overset{?}{=}25,\ 25=25$

**5.** $4(3x-1)=12 \Rightarrow 12x-4=12$

$12x=16 \Rightarrow x=\frac{16}{12}=\frac{4}{3}=1\frac{1}{3}$

check: $4\left(3\cdot\frac{4}{3}-1\right)\overset{?}{=}12,\ 12=12$

**7.** $36=-2(4y-2) \Rightarrow 36=-8y+4$

$-8y=32 \Rightarrow y=-4$

check: $36\overset{?}{=}-2(4(-4)-2),\ 36=36$

**9.** $-5(2x+1)+3x=32 \Rightarrow -10x-5+3x=32$

$-7x=37 \Rightarrow x=-\frac{37}{7}$

check: $-5\left(2\left(-\frac{37}{7}\right)+1\right)+3\left(-\frac{37}{7}\right)\overset{?}{=}32$

$32=32$

**11.** $-2(5y-1)+7y=-1$

$-10y+2+7y=-1 \Rightarrow -3y=-3 \Rightarrow y=1$

check: $-2(5(1)-1)+7(1)\overset{?}{=}-1,\ -1=-1$

**13.** $-4(2x+1)=-5-4x \Rightarrow -8x-4=-5-4x$

$-4x=-1 \Rightarrow x=\frac{1}{4}$

check: $-4\left(2\cdot\frac{1}{4}+1\right)\overset{?}{=}-5-4\cdot\frac{1}{4},\ -6=-6$

**15.** $3(y-4)+6(y+1)=57$

$3y-12+6y+6=57 \Rightarrow 9y=63 \Rightarrow y=7$

check: $3(7-4)+6(7+1)\overset{?}{=}57,\ 57=57$

**17.** $2(x-1)+4(x+2)=18$

$2x-2+4x+8=18 \Rightarrow 6x+6=18$

$6x=12 \Rightarrow x=2$

check: $2(2-1)+4(2+2)\overset{?}{=}18,\ 18=18$

**19.** $10=-2(x-2)+6(x-1)$

$10=-2x+4+6x-6 \Rightarrow 10=4x-2$

$4x=12 \Rightarrow x=3$

check: $10\overset{?}{=}-2(3-2)+6(3-1),\ 10=10$

**21.** $6(x-4)=-9(x+1)+10$

$6x-24=-9x-9+10$

$15x=25 \Rightarrow x=\frac{25}{15}=\frac{5}{3}$

check: $6\left(\frac{5}{3}\right)-24\overset{?}{=}-9\left(\frac{5}{3}+1\right)+10$

$-14=-14$

**23.** $2(y+3)=-6(y-1)-7$

$2y+6=-6y+6-7$

$8y=-7 \Rightarrow y=-\frac{7}{8}$

check: $2\left(-\frac{7}{8}+3\right)\overset{?}{=}-6\left(-\frac{7}{8}-1\right)-7$

$\frac{17}{4}=\frac{17}{4}$

**25.** $(6x^2+4x-1)-(6x^2+9)=12$

$6x^2+4x-1-6x^2-9=12$

$4x-10=12 \Rightarrow 4x=22 \Rightarrow x=\frac{22}{4}=\frac{11}{2}$

check: $(6\cdot\frac{11}{2}^2+4\cdot\frac{11}{2}+1)$

$-(6\cdot\frac{11}{2}^2-9)\overset{?}{=}12,\ 12=12$

**27.** $(5x^2+2x+3)-(5x^2+9)=4x+1$

$5x^2+2x+3-5x^2-9=4x+1$

$2x-6=4x+1 \Rightarrow 2x=-7 \Rightarrow x=-\frac{7}{2}$

check: $\left(5\left(-\frac{7}{2}\right)^2+2\left(-\frac{7}{2}\right)+3\right)$

$-\left(5\left(-\frac{7}{2}\right)^2+9\right)\overset{?}{=}4\left(-\frac{7}{2}\right)+1,\ -13=-13$

**29.** $6(x-1)+2(x+1)=10-3x$

$6x-6+2x+2=10-3x$

$8x-4=10-3x \Rightarrow 11x=14 \Rightarrow x=\frac{14}{11}$

check: $6\left(\frac{14}{11}-1\right)+2\left(\frac{14}{11}+1\right)\overset{?}{=}10-3\cdot\frac{14}{11}$

$\frac{68}{11}=\frac{68}{11}$

**31.** $4(-6x+2)-6x=68$

$-24x+8-6x=68$

$-30x=60 \Rightarrow x=-2$

check: $4(-6(-2)+2)-6(-2)\overset{?}{=}68$

$68=68$

**33.** $(4x^2+3x+1)-(4x^2-2)=5x-2$

$4x^2+3x+1-4x^2+2=5x-2$

$3x+3=5x-2 \Rightarrow -2x=-5 \Rightarrow x=\frac{5}{2}$

**33.** check: $4\left(\frac{5}{2}\right)^2+3\left(\frac{5}{2}\right)+1$

$-\left(4\left(\frac{5}{2}\right)^2-2\right)\stackrel{?}{=}5\left(\frac{5}{2}\right)-2,\ \frac{21}{2}=\frac{21}{2}$

**35.** $2x^2-6x-3-x^2-4=x^2+6$

$x^2-6x-7=x^2+6 \Rightarrow -6x-7=6$

$-6x=13 \Rightarrow x=-\frac{13}{6}$

check: $2\left(-\frac{13}{6}\right)^2-4\stackrel{?}{=}\left(-\frac{13}{6}\right)^2+6,\ \frac{385}{36}=\frac{385}{36}$

**37.** $(x^2+3x-1)-(2x+1)=x^2-5$

$x^2+3x-1-2x-1=x^2-5$

$x-2=-5 \Rightarrow x=-3$

check: $(-3)^2+3(-3)-1$

$-(2(-3)+1)\stackrel{?}{=}(-3)^2-5,\ 4=4$

**39.** $(2x+9)+(3x-2)-(5x+1)$

$=2(x-6)-(x-1)$

$2x+9+3x-2-5x-1=2x-12-x+1$

$6=x-11 \Rightarrow x=17 \neq 19$

19 is not a solution

### Cumulative Review

**41.** $3=3,\ 4=2^2,\ 2=2,\ \text{LCD}=2^2\cdot 3=12$

**43.** $2x=2x,\ x=x,\ \text{LCD}=2x$

**45.** $\frac{2}{5}\cdot 925=370$ pages

### How Am I Doing in 7.1-7.3?

**1.** $-12=a+3,\ a=-12-3=-15$

check: $-12\stackrel{?}{=}-15+3,\ -12=-12$

**2.** $\frac{x}{-2}=4+5^2=4+25=29$

$x=-2(29)=-58$

check: $\frac{-58}{-2}\stackrel{?}{=}4+5^2,\ 29=29$

**3.** $-x=4 \Rightarrow x=-4$

check: $-(-4)\stackrel{?}{=}4,\ 4=4$

**4.** $\frac{21}{-7}=9x+3(-4x)=9x-12x=-3x$

$x=\frac{21}{-7(-3)}=1$

check: $\frac{21}{-7}\stackrel{?}{=}9(1)+3(-4\cdot 1),\ -3=-3$

**5.** $4x+10=14 \Rightarrow 4x=4 \Rightarrow x=1$

check: $4(1)+10\stackrel{?}{=}14,\ 14=14$

**6.** $-5y-9=24,\ -5y=33,\ y=-\frac{33}{5}$

check: $-5\left(-\frac{33}{5}\right)-9\stackrel{?}{=}24,\ 24=24$

**7.** $-3=7-2y,\ -10=-2y,\ y=5$

check: $-3\stackrel{?}{=}7-2(5),\ -3=-3$

**8.** $6x-4=14x-9,\ 5=8x,\ x=\frac{5}{8}$

check: $6\cdot\frac{5}{8}-4\stackrel{?}{=}14\cdot\frac{5}{8}-9,\ -\frac{1}{4}=-\frac{1}{4}$

**9.** $15y-6+5y=8y-1 \Rightarrow 20y=8y+5$

$12y=5 \Rightarrow y=\frac{5}{12}$

check: $15\left(\frac{5}{12}\right)-6+5\left(\frac{5}{12}\right)\stackrel{?}{=}8\left(\frac{5}{12}\right)-1$

$\frac{7}{3}=\frac{7}{3}$

**10.** $7-8=2x-5+5x \Rightarrow 7x-5=-1$

$7x=4 \Rightarrow x=\frac{4}{7}$

check: $7-8\stackrel{?}{=}2\cdot\frac{4}{7}-5+5\cdot\frac{4}{7},\ -1=-1$

**11.** $-3(2x+1)+4x=11 \Rightarrow -6x-3+4x=11$

$-2x=14 \Rightarrow x=-7$

check: $-3(2(-7)+1)+4(-7)\stackrel{?}{=}11,\ 11=11$

**12.** $2(x+1)=-3(x+2)+10$

$2x+2=-3x-6+10 \Rightarrow 5x=2 \Rightarrow x=\frac{2}{5}$

check: $2\left(\frac{2}{5}+1\right)\stackrel{?}{=}-3\left(\frac{2}{5}+2\right)+10,\ \frac{14}{5}=\frac{14}{5}$

**13.** $(3y^2+8y-1)-(3y^2+2)=5$

$3y^2+8y-1-3y^2-2=5$

$8y=8 \Rightarrow y=1$

check: $(3\cdot1^2+8\cdot1-1)-(3\cdot1^2+2)\stackrel{?}{=}5$

$5=5$

### 7.4 Exercises

**1.** To solve $\frac{x}{3}+\frac{x}{7}=10$ we multiply each term by <u>21</u>, so that we clear the fraction.

**3.**

$$\frac{x}{4}+\frac{x}{3}=7$$

$$\boxed{12}\cdot\frac{x}{4}+\boxed{12}\frac{x}{3}=\boxed{12}\cdot 7$$

$$\boxed{3}x+\boxed{4}x=84$$

$$\boxed{7}x=84$$

$$x=\boxed{12}$$

**5.** $\frac{x}{2}\cdot4+\frac{x}{4}\cdot4=12\cdot4 \Rightarrow 2x+x=48$

$3x=48 \Rightarrow x=16$

check: $\frac{16}{2}+\frac{16}{4}\stackrel{?}{=}12,\ 12=12$

**7.** $\frac{x}{6}\cdot12+\frac{x}{4}\cdot12=5\cdot12 \Rightarrow 2x+3x=60$

$5x=60 \Rightarrow x=12$

check: $\frac{12}{6}+\frac{12}{4}\stackrel{?}{=}5,\ 5=5$

**9.** $\frac{x}{2}\cdot14-\frac{x}{7}\cdot14=10\cdot14 \Rightarrow 7x-2x=140$

$5x=140 \Rightarrow x=28$

check: $\frac{28}{2}-\frac{28}{7}\stackrel{?}{=}10,\ 10=10$

**11.** $3x\cdot6+\frac{2}{3}\cdot6=\frac{9}{2}\cdot6 \Rightarrow 18x+4=27$

$18x=23 \Rightarrow x=\frac{23}{18}$

check: $3\cdot\frac{23}{18}+\frac{2}{3}\stackrel{?}{=}\frac{9}{2},\ \frac{9}{2}=\frac{9}{2}$

**13.** $5x\cdot8+\frac{1}{8}\cdot8=\frac{3}{4}\cdot8 \Rightarrow 40x+1=6$

$40x=5 \Rightarrow x=\frac{5}{40}=\frac{1}{8}$

**13.** check: $5\cdot\frac{1}{8}+\frac{1}{8}\stackrel{?}{=}\frac{3}{4},\ \frac{3}{4}=\frac{3}{4}$

**15.** $5x\cdot 8-\frac{1}{2}\cdot 8=\frac{1}{8}\cdot 8\Rightarrow 40x-4=1$

$40x=5\Rightarrow x=\frac{5}{40}=\frac{1}{8}$

check: $5\cdot\frac{1}{8}-\frac{1}{2}\stackrel{?}{=}\frac{1}{8},\ \frac{1}{8}=\frac{1}{8}$

**17.** $-2x\cdot 14+\frac{1}{2}\cdot 14=\frac{3}{7}\cdot 14\Rightarrow -28x+7=6$

$-28x=-1\Rightarrow x=\frac{-1}{-28}=\frac{1}{28}$

check: $-2\cdot\frac{1}{28}+\frac{1}{2}\stackrel{?}{=}\frac{3}{7},\ \frac{3}{7}=\frac{3}{7}$

**19.** $-4x\cdot 6+\frac{2}{3}\cdot 6=\frac{1}{6}\cdot 6\Rightarrow -24x+4=1$

$-24x=-3\Rightarrow x=\frac{-3}{-24}=\frac{1}{8}$

check: $-4\cdot\frac{1}{8}+\frac{2}{3}\stackrel{?}{=}\frac{1}{6},\ \frac{1}{6}=\frac{1}{6}$

**21.** $\frac{x}{5}+x=6\Rightarrow\frac{6x}{5}=6\Rightarrow 6x=30\Rightarrow x=5$

check: $\frac{5}{5}+5\stackrel{?}{=}6,\ 6=6$

**23.** $\frac{x}{3}+x=6\Rightarrow\frac{4x}{3}=6\Rightarrow 4x=18\Rightarrow x=\frac{9}{2}$

check: $\frac{\frac{9}{2}}{3}+\frac{9}{2}\stackrel{?}{=}6,\ 6=6$

**25.** $\frac{x}{2}+x=3\Rightarrow x+2x=6\Rightarrow 3x=6\Rightarrow x=2$

check: $\frac{2}{2}+2\stackrel{?}{=}3,\ 3=3$

**27.** $\frac{x}{4}-3x=3\Rightarrow x-12x=12$

$-11x=12\Rightarrow x=-\frac{12}{11}$

check: $\frac{-\frac{12}{11}}{4}-3\left(-\frac{12}{11}\right)\stackrel{?}{=}3,\ 3=3$

**29.** $2x+\frac{1}{3}=\frac{1}{6}\Rightarrow 12x+2=1$

$12x=-1\Rightarrow x=-\frac{1}{12}$

check: $2\cdot\frac{-1}{12}+\frac{1}{3}\stackrel{?}{=}\frac{1}{6},\ \frac{1}{6}=\frac{1}{6}$

**31.** $\frac{x}{2}+x=6\Rightarrow x+2x=12$

$3x=12\Rightarrow x=4$

check: $\frac{4}{2}+4\stackrel{?}{=}6,\ 6=6$

**33.** $\frac{x}{2}+\frac{x}{5}=7,\ 5x+2x=70,\ 7x=70,\ x=10$

check: $\frac{10}{2}+\frac{10}{5}\stackrel{?}{=}7,\ 7=7$

**35.** $\frac{5}{2}\cdot 6+\frac{x}{3}\cdot 6=\frac{1}{6}\cdot 6\Rightarrow 15+2x=1$

$2x=-14\Rightarrow x=-7$

check: $\frac{5}{2}+\frac{-7}{3}\stackrel{?}{=}\frac{1}{6},\ \frac{1}{6}=\frac{1}{6}$

**37.** $\frac{3}{2}\cdot 10+\frac{x}{10}\cdot 10=\frac{1}{5}\cdot 10\Rightarrow 15+x=2$

$x=-13$

check: $\frac{3}{2}+\frac{-13}{10}\stackrel{?}{=}\frac{1}{5},\ \frac{1}{5}=\frac{1}{5}$

**39.** $\frac{x}{2}\cdot 6-\frac{2}{6}\cdot 6=-\frac{5}{6}\cdot 6 \Rightarrow 3x-2=-5$

$3x=-3 \Rightarrow x=-1$

check: $\frac{-1}{2}-\frac{2}{6}\overset{?}{=}-\frac{5}{6},\quad -\frac{5}{6}=-\frac{5}{6}$

**41.** $x+\frac{2}{3}+2=\frac{3}{2}+\frac{1}{4}$

$12x+8+24=18+3 \Rightarrow 12x+32=21$

$12x=-11 \Rightarrow x=-\frac{11}{12}$

**43.** $4+\frac{6}{x}+\frac{2}{5}=\frac{3}{2x} \Rightarrow 40x+60+4x=15$

$44x=-45 \Rightarrow x=-\frac{45}{44}$

**45.** $\frac{1}{3}\left(\frac{x}{2}+3\right)+\frac{1}{4}=\frac{5}{6} \Rightarrow 4\left(\frac{x}{2}+3\right)+3=10$

$2x+12+3=10 \Rightarrow 2x=-5 \Rightarrow x=-\frac{5}{2}$

## Cumulative Review

**47.** Six more than twice a number: $6+2x$

**49.** The sum of 4 and $x$: $4+x$

**51.** $1-\frac{1}{3}-\frac{1}{4}=\frac{12-4-3}{12}=\frac{5}{12}$ of restraurant

## 7.5 Exercises

**1.** (a) $2n+5=15$

(b) $2n+5=15 \Rightarrow 2n=10 \Rightarrow n=5$

**3.** (a) $3n-4=5$

(b) $3n-4=5 \Rightarrow 3n=9 \Rightarrow n=3$

**5.** (a) $2(5+n)=12$

(b) $2(5+n)=12 \Rightarrow 10+2n=12$

$2n=2 \Rightarrow n=1$

**7.** $P=60=2L+2W=2(x+15)+2(12)$

$2x+30+24=60 \Rightarrow 2x=6 \Rightarrow x=3$ m

**9.** $P=x+(x+2)+(x+1)=12$

$3x+3=12 \Rightarrow 3x=9 \Rightarrow x=3$

First side $=x=3$ cm

second side $=x+1=4$ cm

third side $=x+2=5$ cm

**11.** $A=90=LW=2x(15) \Rightarrow 30x=90$

$x=3$ in.

**13.** (a) $P=70=2(15)+2(x+12)$

$2x+24+30=70 \Rightarrow 2x=16 \Rightarrow x=8$ ft

(b) width $=x+12=8+12=20$ ft

**15.** (a) $A=(9+x)(7)=105 \Rightarrow 63+7x=105$

$7x=42 \Rightarrow x=6$ ft

(b) $L=9+x=9+6=15$ ft

**17.** (a) $J$ = Jerome's salary

$J-7500$ = cashier's salary

(b) $J+(J-7500)=62{,}000$

(c) $2J-7500=62{,}000 \Rightarrow 2J=69{,}500$

$J=\$34{,}750$, Jerome's salary

$J-7500=\$27{,}250$, cashier's salary

(d) $34{,}750+(34{,}750-7500)\overset{?}{=}62{,}000$

$62{,}000=62{,}000$

**19.** (a) $x$ = miles he drove the second day

$x+85$ = miles he drove the first day

(b) $x+(x+85)=385$

(c) $x+(x+85)=385 \Rightarrow 2x=300$

$x=150$ mi the second day

$x+85=235$ mi the first day

(d) $150+(150+85)\overset{?}{=}385,\ 385=385$

**21.** (a) $s =$ flight time of second flight

$\frac{1}{2} \cdot s =$ flight time of first flight

(b) $s + \left(\frac{1}{2} \cdot s\right) = 15$

(c) $s + \left(\frac{1}{2} \cdot s\right) = 15 \Rightarrow \frac{3}{2} \cdot s = 15$

$s = 10$ hr, time of second flight

$\frac{1}{2} \cdot s = 5$ hr, time of first flight

(d) $10 + \frac{10}{2} \stackrel{?}{=} 15,\ 15 = 15$

**23.** (a) $L =$ first side of triangle

$2L =$ second side of triangle

$L + 12 =$ third side of triangle

(b) $L + 2L + (L + 12) = 120$

(c) $4L = 108 \Rightarrow L = 27$ m, first side

$2L = 54$ m, second side

$L + 12 = 39$ m, third side

(d) $27 + 2(27) + (27 + 12) \stackrel{?}{=} 120$

$120 = 120$

**25.** (a) $W =$ width, $3W - 2 =$ length

(b) $2W + 2(3W - 2) = 68$

(c) $2W + 6W - 4 = 68 \Rightarrow 8W = 72$

$W = 9$ m, width

$3W - 2 = 25$ m, length

(d) $2(9) + 2(3(9) - 2) \stackrel{?}{=} 68,\ 68 = 68$

**27.** (a) $x =$ fall students

$95 + x =$ spring students

$x - 75 =$ summer students

(b) $x + (95 + x) + (x - 75) = 395$

(c) $3x + 20 = 395,\ 3x = 375$

$x = 125$ fall students

$95 + x = 220$ spring students

$x - 75 = 50$ summer students

(d) $125 + (95 + 125) + (125 - 75) \stackrel{?}{=} 395$

$395 = 395$

**29.** $3x - 1 + 5x - 3 = 180,\ 8x - 4 = 180$

$8x = 184,\ x = 23$

$3x - 1 = 68°,\ 5x - 3 = 112°$

**31.** $4s = s^2,\ s^2 - 4s = 0,\ s(s - 4) = 0$

$s = 4$ units

## Cumulative Review

**33.** $3044 \div 21 = 144$ R20

**35.** $A = bh = 3(4) = 12\ \text{ft}^2$

**37.** $L = 13\frac{1}{2}$ in.

$W = \frac{1}{2} \cdot 13\frac{1}{2} = 6\frac{3}{4}$ in.

$H = 2 \cdot \frac{5}{8} = 1\frac{1}{4}$ in.

## Putting Your Skills to Work, Problems for Individual Analysis

**1.** $C = 750 = 2x + 50 \Rightarrow 2x = 700$

$x = 350$ hinges

**2.** $R = 3x - 5 = 3(350) - 5 = \$1045$

**3.** $P = R - C = 1045 - 750 = \$295$

**4.** $3x - 5 = 2x + 50 \Rightarrow x = 55$ hinges

**5.** $P = 500 = (3x-5)-(2x+50)$
$x-55 = 500 \Rightarrow x = 555$ hinges

## Putting Your Skills to Work, Problems for Group Investigation and Study

**1.** $C = 2x+50+100 = 2x+150$

**2.** $R = 2(3x-5) = 6x-10$

**3.** $P = R-C = (6x-10)-(2x+150)$

**4.** (a) $6x-10 = 2x+150 \Rightarrow 4x = 160$
$x = 40$ hinges
(b) $40 < 55$, less

(c) Answers may vary. One explanation: since the break-even point is less for Model B, the company will begin to show a profit sooner.

**5.**

| Model A | | Model B | |
|---|---|---|---|
| $x$ hinges | Profit | $x$ hinges | Profit |
| 15 | –\$40 | 15 | –\$100 |
| 30 | –25 | 30 | –40 |
| 40 | –15 | 40 | 0 |
| 45 | –10 | 45 | 20 |
| 55 | 0 | 55 | 60 |

**6.** (a) There is a loss.
(b) Model B is more profitable because producing 45 Model B yields a profit of \$20 while 45 Model A hinges yields a loss of \$10.

**7.** (a) Model A has 0 profit for 55 hinges. Model B has0 profit for 40 hinges.
(b) Yes. The break-even point occurs when the profit is zero.

## Chapter 7 Review Problems

**1.** $x-6 = 48 \Rightarrow x = 48+6 = 54$

check: $54-6 \stackrel{?}{=} 48,\ 48 = 48$

**2.** $x+9 = -56 \Rightarrow x = -56-9 = -65$

check: $-65+9 \stackrel{?}{=} -56,\ -56 = -56$

**3.** $-13-20 = 3x+15-2x$
$x+15 = -33 \Rightarrow x = -48$

check: $-13-20 \stackrel{?}{=} 3(-48)+15-2(-48)$
$-33 = -33$

**4.** $\frac{y}{-2} = -8+3^2 = -8+9 = 1 \Rightarrow y = -2(1)$
$y = -2$

check: $\frac{-2}{-2} \stackrel{?}{=} -8+3^2,\ 1 = 1$

**5.** $-6+2^2 = \frac{x}{-5} \Rightarrow \frac{x}{-5} = -6+4 = -2$
$x = -5(-2) = 10$

check: $-6+2^2 \stackrel{?}{=} \frac{10}{-5},\ -2 = -2$

**6.** $\frac{-15}{3} = -2(5x)-6x,\ -16x = -5,\ x = \frac{5}{16}$

check: $\frac{-15}{3} \stackrel{?}{=} -2\left(5\left(\frac{5}{16}\right)\right)-6\left(\frac{5}{16}\right),\ -5 = -5$

**7.** $-3x+2(4x) = \frac{18}{-2} \Rightarrow -3x+8x = -9$

$5x = -9 \Rightarrow x = \frac{-9}{5}$

check: $-3\left(\frac{-9}{5}\right)+2\left(4\cdot\frac{-9}{5}\right) \stackrel{?}{=} \frac{18}{-2}$
$-9 = -9$

**8.** $-x = 42 \Rightarrow x = -42$

check: $-(-42) \stackrel{?}{=} 42,\ 42 = 42$

**9.** $-y = \frac{1}{2} \Rightarrow y = -\frac{1}{2}$

check: $-\left(-\frac{1}{2}\right) \stackrel{?}{=} \frac{1}{2}, \ \frac{1}{2} = \frac{1}{2}$

**10.** $6x - 8 = 34 \Rightarrow 6x = 42 \Rightarrow x = 7$

check: $6(7) - 8 \stackrel{?}{=} 34, \ 34 = 34$

**11.** $-2y + 6 = 58 \Rightarrow -2y = 52 \Rightarrow y = -26$

check: $-2(-26) + 6 \stackrel{?}{=} 58, \ 58 = 58$

**12.** $-20 = 8 - 7y \Rightarrow -7y = -28 \Rightarrow y = 4$

check: $-20 \stackrel{?}{=} 8 - 7(4), \ -20 = -20$

**13.** $46 = 10 - 4x \Rightarrow 4x = -36 \Rightarrow x = -9$

check: $46 \stackrel{?}{=} 10 - 4(-9), \ 46 = 46$

**14.** $1 - 4y = 19 \Rightarrow 4y = -18$

$y = \frac{-18}{4} = \frac{-9}{2}$

check: $1 - 4\left(\frac{-9}{2}\right) \stackrel{?}{=} 19, \ 19 = 19$

**15.** $8x - 7 - 5x = 15 \Rightarrow 3x = 22 \Rightarrow x = \frac{22}{3}$

check: $8\left(\frac{22}{3}\right) - 7 - 5\left(\frac{22}{3}\right) \stackrel{?}{=} 15, \ 15 = 15$

**16.** $-6x = 9x + 36 \Rightarrow -15x = 36$

$x = \frac{36}{-15} = -\frac{12}{5}$

check: $-6\left(-\frac{12}{5}\right) \stackrel{?}{=} 9\left(-\frac{12}{5}\right) + 36$

$\frac{72}{5} = \frac{72}{5}$

**17.** $5x = 3x + 30 \Rightarrow 2x = 30 \Rightarrow x = 15$

check: $5(15) \stackrel{?}{=} 3(15) + 30, \ 75 = 75$

**18.** $3x - 8 = 5x + 6 \Rightarrow 2x = -14 \Rightarrow x = -7$

check: $3(-7) - 8 \stackrel{?}{=} 5(-7) + 6, \ -29 = -29$

**19.** $-3 + 2y + 6 = -4y + 12$

$2y + 3 = -4y + 12 \Rightarrow 6y = 9 \Rightarrow y = \frac{9}{6} = \frac{3}{2}$

check: $-3 + 2\left(\frac{3}{2}\right) + 6 \stackrel{?}{=} -4\left(\frac{3}{2}\right) + 12$

$6 = 6$

**20.** $6y - 8 + 2y = -6 + 9y$

$8y - 8 = -6 + 9y \Rightarrow y = -2$

check: $6(-2) - 8 + 2(-2) \stackrel{?}{=} -6 + 9(-2)$

$-24 = -24$

**21.** $8x - 9 - 5x + 18 = -3x - 2 + 2x$

$3x + 9 = -x - 2 \Rightarrow 4x = -11 \Rightarrow x = \frac{-11}{4}$

check : $8\left(\frac{-11}{4}\right) - 9 - 5\left(\frac{-11}{4}\right) + 18$

$\stackrel{?}{=} -3\left(\frac{-11}{4}\right) - 2 + 2\left(\frac{-11}{4}\right), \ \frac{3}{4} = \frac{3}{4}$

**22.** $-3x - 2 - 9x + 5 = 2x + 8 - 7x$

$-12x + 3 = -5x + 8$

$-7x = 5 \Rightarrow x = -\frac{5}{7}$

check: $-3\left(-\frac{5}{7}\right) - 2 - 9\left(-\frac{5}{7}\right) + 5$

$\stackrel{?}{=} 2\left(-\frac{5}{7}\right) + 8 - 7\left(-\frac{5}{7}\right)$

$\frac{81}{7} = \frac{81}{7}$

**23.** $-3(x+5)=21 \Rightarrow -3x-15=21$

$-3x=36 \Rightarrow x=-12$

check: $-3(-12+5)\stackrel{?}{=}21,\ 21=21$

**24.** $4(2x+9)+5x=-3 \Rightarrow 8x+36+5x=-3$

$13x=-39 \Rightarrow x=-3$

check: $4(2(-3)+9)+5(-3)\stackrel{?}{=}-3, -3=-3$

**25.** $7(x+1)+3(x+1)=-10$

$7x+7+3x+3=-10 \Rightarrow 10x+10=-10$

$10x=-20 \Rightarrow x=-2$

check: $7(-2+1)+3(-2+1)\stackrel{?}{=}-10, -10=-10$

**26.** $-3(x-7)=5(x+6)-10$

$-3x+21=5x+30-10=5x+20$

$-8x=-1 \Rightarrow x=\dfrac{1}{8}$

check: $-3\left(\dfrac{1}{8}-7\right)\stackrel{?}{=}5\left(\dfrac{1}{8}+6\right)-10,\ \dfrac{165}{8}=\dfrac{165}{8}$

**27.** $(9y^2+8y-2)-(9y^2+5y-1)=14$

$9y^2+8y-2-9y^2-5y+1=14$

$3y-1=14 \Rightarrow 3y=15 \Rightarrow y=5$

check: $9(5)^2+8(5)-2-9(5)^2-5(5)+1\stackrel{?}{=}14$

$14=14$

**28.** $(5x^2+x-2)-(5x^2-5)=6x+9$

$5x^2+x-2-5x^2+5=6x+9$

$x+3=6x+9 \Rightarrow -5x=6 \Rightarrow x=-\dfrac{6}{5}$

check: $5\left(-\dfrac{6}{5}\right)^2-\dfrac{6}{5}-2$

$-5\left(-\dfrac{6}{5}\right)^2+5\stackrel{?}{=}6\left(-\dfrac{6}{5}\right)+9,\ \dfrac{9}{5}=\dfrac{9}{5}$

**29.** $\dfrac{x}{3}\cdot 12+\dfrac{x}{4}\cdot 12=7\cdot 12 \Rightarrow 4x+3x=84$

$7x=84 \Rightarrow x=12$

check: $\dfrac{12}{3}\cdot 12+\dfrac{12}{4}\cdot 12=7\cdot 12,\ 84=84$

**30.** $\dfrac{x}{5}-\dfrac{x}{2}=6 \Rightarrow 2x-5x=60$

$-3x=60 \Rightarrow x=-20$

check: $\dfrac{-20}{5}-\dfrac{-20}{2}\stackrel{?}{=}6,\ 6=6$

**31.** $y+\dfrac{y}{9}=-10 \Rightarrow 9y+y=-90$

$10y=-90 \Rightarrow y=-9$

check: $-9+\dfrac{-9}{9}=-10,\ -10=-10$

**32.** $2y+\dfrac{y}{-4}=7 \Rightarrow -8y+y=-28$

$-7y=-28 \Rightarrow y=4$

check: $2(4)+\dfrac{4}{-4}\stackrel{?}{=}7,\ 7=7$

**33.** $2x-\dfrac{3}{4}=\dfrac{1}{3} \Rightarrow 24x-9=4$

$24x=13 \Rightarrow x=\dfrac{13}{24}$

check: $2\cdot\dfrac{13}{24}-\dfrac{3}{4}\stackrel{?}{=}\dfrac{1}{3},\ \dfrac{1}{3}=\dfrac{1}{3}$

**34.** $2x-\dfrac{3}{4}=\dfrac{1}{2} \Rightarrow 8x-3=2$

$8x=5 \Rightarrow x=\dfrac{5}{8}$

check: $2\cdot\dfrac{5}{8}-\dfrac{3}{4}\stackrel{?}{=}\dfrac{1}{2},\ \dfrac{1}{2}=\dfrac{1}{2}$

**35.** $-\frac{1}{3}=3y+\frac{1}{2}\Rightarrow -2=18y+3$

$18y=-5\Rightarrow y=-\frac{5}{18}$

check: $-\frac{1}{3}\stackrel{?}{=}3\left(-\frac{5}{18}\right)+\frac{1}{2},\ -\frac{1}{3}=-\frac{1}{3}$

**36.** (a) $2n+4=16$
(b) $2n=12\Rightarrow n=6$

**37.** (a) $2(16)+2W=54$
(b) $32+2W=54\Rightarrow 2W=22$
$W=11$ ft

**38.** (a) $P=56=2(15+x)+2(8)$
$30+2x+16=56\Rightarrow 2x=10\Rightarrow x=5$ ft
(b) $L=15+x=15+5=20$ ft

**39.** (a) $J=$ Jamie's salary
$J-8500=$ cashier's salary
(b) $J+(J-8500)=59,000$
(c) $2J=67,500$
$J=\$33,750$, Jamie's salary
$J-8500=\$25,250$, cashier's salary
(d) $33,750+(33,750-8500)\stackrel{?}{=}59,000$
$59,000=59,000$

**40.** (a) $L=$ fall students
$L+85=$ spring students
$L-65=$ summer students
(b) $L+(L+85)+(L-65)=491$
(c) $3L+20=491\Rightarrow 3L=471$
$L=157$ fall students
$L+85=242$ spring students
$L-65=92$ summer students
(d) $157-(157+85)+(157-65)\stackrel{?}{=}491$
$491=491$

## Chapter 7 Test

**1.** $-6x=72\Rightarrow x=\frac{72}{-6}=-12$

check: $-6(-12)\stackrel{?}{=}72,\ 72=72$

**2.** $7x+3-6x=-4\Rightarrow x=-4-3=-7$

check: $7(-7)+3-6(-7)\stackrel{?}{=}-4,\ -4=-4$

**3.** $\frac{a}{-3}=-4+6=2\Rightarrow a=-3(2)=-6$

check: $\frac{-6}{-3}\stackrel{?}{=}-4+6,\ 2=2$

**4.** $5(2x)=-30\Rightarrow 10x=-30\Rightarrow x=-3$

check: $5(2(-3))\stackrel{?}{=}-30,\ -30=-30$

**5.** $5x-11=-1\Rightarrow 5x=10\Rightarrow x=2$

check: $5(2)-11\stackrel{?}{=}-1,\ -1=-1$

**6.** $\frac{x}{5}=-2+4^2=-2+16=14$
$x=5(14)=70$

check: $\frac{70}{5}\stackrel{?}{=}-2+4^2,\ 14=14$

**7.** $4(-2y)+5y=12\Rightarrow -8y+5y=12$
$-3y=12\Rightarrow y=-4$

check: $4(-2(-4))+5(-4)\stackrel{?}{=}12,\ 12=12$

**8.** $12y+6-11y-1=-8+10=2$
$y+5=2\Rightarrow y=-3$

check: $12(-3)+6-11(-3)-1\stackrel{?}{=}-8+10$
$2=2$

**9.** $-3 = 7 - 4y \Rightarrow 4y = 10 \Rightarrow y = \frac{5}{2}$

check: $-3 \stackrel{?}{=} 7 - 4\left(\frac{5}{2}\right),\ -3 = -3$

**10.** $6x + 4 - 9x = 15 \Rightarrow -3x = 11 \Rightarrow x = -\frac{11}{3}$

check: $6\left(-\frac{11}{3}\right) + 4 - 9\left(-\frac{11}{3}\right) \stackrel{?}{=} 15$

$15 = 15$

**11.** $14x = -2x + 16 \Rightarrow 16x = 16 \Rightarrow x = 1$

check: $14(1) \stackrel{?}{=} -2(1) + 16,\ 14 = 14$

**12.** $2x + 4(3x - 6) = 4 - (6x + 2)$

$2x + 12x - 24 = 4 - 6x - 2$

$14x - 24 = -6x + 2 \Rightarrow 20x = 26$

$x = \frac{26}{20} = \frac{13}{10}$

check: $2\left(\frac{13}{10}\right) + 4\left(3\left(\frac{13}{10}\right) - 6\right)$

$\stackrel{?}{=} 4 - \left(6\left(\frac{13}{10}\right) + 2\right)$

$-\frac{29}{5} = -\frac{29}{5}$

**13.** $3(2x + 6) + 3x = -27$

$6x + 18 + 3x = -27 \Rightarrow 9x = -45 \Rightarrow x = -5$

**14.** $4(x - 1) = -6(x + 2) + 48$

$4x - 4 = -6x - 12 + 48 = -6x + 36$

$10x = 40 \Rightarrow x = 4$

**15.** $(3x^2 - 2x + 1) + (-3x^2 - 10) = 5x + 5$

$3x^2 - 2x + 1 - 3x^2 - 10 = 5x + 5$

$-2x - 9 = 5x + 5 \Rightarrow 7x = -14 \Rightarrow x = -2$

**16.** $\frac{x}{5} + \frac{x}{2} = 7 \Rightarrow 2x + 5x = 70$

$7x = 70$

$x = 10$

**17.** $2x + \frac{1}{3} = \frac{1}{2} \Rightarrow 12x + 4 = 3$

$12x = -1$

$x = -\frac{1}{12}$

**18.** $\frac{x}{6} + x = 14 \Rightarrow \frac{7}{6}x = \overset{2}{\cancel{14}} \Rightarrow x = 6(2) = 12$

**19.** $s =$ length of second side

$s + 2 =$ length of first side

$3s =$ length of third side

**20.** $s + (s + 2) + 3s = 42$

**21.** $5s + 2 = 42 \Rightarrow 5s = 40 \Rightarrow s = 8$

$s = 8$ ft, second side

$s + 2 = 10$ ft, first side

$3s = 24$ ft, third side

**22.** $A =$ Anna's annual salary

$A - 4000 =$ clerk's annual salary

**23.** $A + (A - 4000) = 61,200$

**24.** $2A - 4000 = 61,200$

$2A = 65,200$

$A = \$32,600$, Anna's salary

$A - 4000 = \$28,600$, clerk's salary

## Cumulative Test for Chapters 1-7

**1.** $-3 + 4 + (-1) + 8 = 1 + 7 = 8$

**2.** $2-8-6-9=-6-6-9=-12-9=-21$

**3.** $(-8)(-2)(2)(-1)=16(-2)=-32$

**4.** $\frac{16}{2}=8$

**5.** The opposite of 9 is $\underline{-9}$.

**6.** The opposite of $-12$ is $\underline{12}$.

**7.** The absolute value of 9 is $\underline{9}$.

**8.** The absolute value of $-12$ is $\underline{12}$.

**9.** $\frac{15}{2}=7\frac{1}{2}$

**10.** $6\frac{3}{4}=\frac{4\cdot 6+3}{4}=\frac{27}{4}$

**11.** $\frac{25}{40}=\frac{5}{8}$

**12.** $\frac{3}{4}\cdot\frac{1}{4}=\frac{3}{16}$ lb

**13.** $2\cdot\frac{3}{4}=\frac{3}{2}=1\frac{1}{2}$ lb

**14.** $\frac{1}{2}\cdot\frac{3}{4}=\frac{3}{8}$ lb

**15.** $(2x^2+3xy+4)+(-8x^2+3)$
$=2x^2+3xy+4-8x^2+3$
$=-6x^2+3xy+7$

**16.** $(8x^2-6x+1)-(6x^2+4x-10)$
$=8x^2-6x+1-6x^2-4x+10$
$=2x^2-10x+11$

**17.** $12xy+4x^2y-8x=4x(3y+xy-2)$

**18.** $y+10=-2 \Rightarrow y=-2-10=-12$

check: $-12+12\stackrel{?}{=}-2,\ -2=-2$

**19.** $\frac{x}{-3}=2^2+6=4+6=10$

$x=-3(10) \Rightarrow x=-30$

check: $\frac{-30}{-3}\stackrel{?}{=}2^2+6,\ 10=10$

**20.** $-3x+1=-11 \Rightarrow -3x=-12 \Rightarrow x=4$

check: $-3(4)+1\stackrel{?}{=}-11,\ -11=-11$

**21.** $-4(3x)=36 \Rightarrow 3x=-9 \Rightarrow x=-3$

check: $-4(3(-3))\stackrel{?}{=}36,\ 36=36$

**22.** $\frac{x}{3}+\frac{x}{2}=5 \Rightarrow 2x+3x=30$

$5x=30 \Rightarrow x=6$

check: $\frac{6}{3}+\frac{6}{2}\stackrel{?}{=}5,\ 5=5$

**23.** $-3x+\frac{x}{2}=4 \Rightarrow -6x+x=8$

$-5x=8 \Rightarrow x=-\frac{8}{5}$

check: $-3\left(-\frac{8}{5}\right)+\frac{-\frac{8}{5}}{2}\stackrel{?}{=}4,\ 4=4$

**24.** $6x = -8 - 2x + 14 \Rightarrow 8x = 6$

$x = \frac{6}{8} \Rightarrow x = \frac{3}{4}$

check: $6\left(\frac{3}{4}\right) \stackrel{?}{=} -8 - 2\left(\frac{3}{4}\right) + 14,\ \frac{9}{2} = \frac{9}{2}$

**25.** $4(y-1) = 6 - (y+8) + 12$

$4y - 4 = 6 - y - 8 + 12$

$5y = 14 \Rightarrow y = \frac{14}{5}$

check: $4\left(\frac{14}{5} - 1\right) \stackrel{?}{=} 6 - \left(\frac{14}{5} + 8\right) + 12$

$\frac{36}{5} = \frac{36}{5}$

**26.** $s =$ length of first side
$s - 3 =$ length of second side
$2s =$ length of third side

**27.** $s + (s - 3) + 2s = 29$

**28.** $4s - 3 = 29 \Rightarrow 4s = 32 \Rightarrow s = 8$
$s = 8$ in., first side
$s - 3 = 5$ in, second side
$2s = 16$ in., third side

**29.** $J =$ Juan's annual salary
$J - 5500 =$ cashier's annual salary

**30.** $J + (J - 5500) = 52{,}000$

**31.** $2J - 5500 = 52{,}000 \Rightarrow 2J = 57{,}500$
$J = \$28{,}750$, Juan's annual salary
$J - 5500 = \$23{,}250$, cashier's salary

# Chapter 8

## 8.1 Exercises

**1.** When we change 9 to 10, we write 0 and carry 1 to the 2, the next place value to the left. Thus the 2 changes to 3.

**3.** 5.32: five and thirty-two hundredths

**5.** 0.428: Four hundred twenty-eight thousandths

**7.** Three hundred twenty-four thousandths: 0.324

**9.** Fifteen and three hundred forty-six ten thousandths: 15.0346

**11.** Twenty-five and 54/100

**13.** One hundred forty-three and 56/100

**15.** $0.7 = \frac{7}{10}$

**17.** $3.64 = 3\frac{64}{100}$

**19.** $0.1743 = \frac{1743}{10,000}$

**21.** $100.011 = 100\frac{11}{1000}$

**23.** $6\frac{1}{10} = 6.1$

**25.** $12\frac{37}{1000} = 12.037$

**27.** $\frac{2}{100} = 0.02$

**29.** $\frac{1}{1000} = 0.001$

**31.** $0.426 < 0.429$

**33.** $0.63 > 0.62$

**35.** $0.36 < 0.366$

**37.** $0.7431 > 0.743$

**39.** $0.3 > 0.27$

**41.** $0.304 < 0.34$

**43.** $523.7235 = 523.72$, nearest hundredth

**45.** $43.995 = 44.00$, nearest hundredth

**47.** $9.0546 = 9.1$, nearest tenth

**49.** $462.931 = 462.9$, nearest tenth

**51.** $312.95144 = 312.951$, nearest thousandth

**53.** $1286.3496 = 1286.350$, nearest thousandth

**55.** $0.063148 = 0.0631$, nearest ten-thousandth

**57.** $0.047357 = 0.0474$, nearest ten-thousandth

**59.** 42.5 million: 43 million, nearest million
34.6 million: 35 million, nearest million

**61.** \$15.25: \$15, nearest dollar

**63.** $\$14 + \$18 + \$16 = \$48$

**65.** 0.73, $\frac{7}{10}$, 0.071, 0.007, 0.0069

**Cumulative Review**

**67.** $-15-(-6)=-15+6=-9$

**69.** $-45\div 9=-5$

**71.** $\frac{2}{3}\cdot\frac{2}{2}+\frac{1}{2}\cdot\frac{3}{3}=\frac{4+3}{6}=\frac{7}{6}=1\frac{1}{6}$ lb of nuts

**8.2 Exercises**

**1.** To add numbers in decimal notation, we <u>line up</u> the decimal points.

**3.** When subtracting decimals, we place the decimal point in the answer <u>in line</u> with the decimal point in the problem.

**5.** $\begin{array}{r} 0.34 \\ \underline{+5.23} \\ 5.57 \end{array}$

**7.** $\begin{array}{r} 0.23 \\ \underline{+3.46} \\ 3.69 \end{array}$

**9.** $\begin{array}{r} 63.2 \\ \underline{+0.2348} \\ 63.4348 \end{array}$

**11.** $\begin{array}{r} 73.0 \\ 7.54 \\ \underline{+0.483} \\ 81.023 \end{array}$

**13.** $\begin{array}{r} 73.1 \\ \underline{+0.3169} \\ 73.4169 \end{array}$

**15.** $\begin{array}{r} 25.0 \\ 2.73 \\ \underline{+0.423} \\ 28.153 \end{array}$

**17.** $\begin{array}{r} 53.783 \\ \underline{-2.34} \\ 51.443 \end{array}$

**19.** $\begin{array}{r} 616.78 \\ \underline{-\ 3.9} \\ 612.88 \end{array}$

**21.** $\begin{array}{r} 20.0 \\ \underline{-0.16} \\ 19.84 \end{array}$

**23.** $\begin{array}{r} -12.1 \\ \underline{-0.23} \\ -12.33 \end{array}$

**25.** $\begin{array}{r} -91.13 \\ \underline{-14.213} \\ -105.343 \end{array}$

**27.** $\begin{array}{r} -8.69 \\ \underline{-(-4.12)} \\ -4.57 \end{array}$

**29.** $2.3x+3.9x=6.2x$

**31.** $24.8y-11.3y=13.5y$

**33.** $3.5x+9.1x-y=12.6x-y$

**35.** $1.4x+6.2y+3.5x=4.9x+6.2y$

**37.** (a) $-3.4+(-2.1)=-5.5$
(b) $9.7-(-5.4)=15.1$
(c) $-9.2-4.1=-13.3$

**39.** (a) $4.6x+2x=6.6x$
(b) $3.04y-7.5y=-4.46y$
(c) $x-0.25x=0.75x$

**41.** $y-0.861=9-0.861=8.139$

**43.** $211.2-n=211.2-9.72=201.48$

**45.** $x+2.3=-6.7+2.3=-4.4$

**47.** estimate: $1+2+1=4$ in.

**49.** estimate: $1763-162-61-48=\$1492$

**51.** $100-72.31=\$27.69$

**53.** $10.75-10.54=0.21$ sec faster

**55.** $11.40-10.54=0.86$ sec

**57.** $-2.3-(-0.24)+4.6-9$
$=-2.3+0.24+4.6-9$
$=-2.06-4.4$
$=-6.46$

**59.** $\frac{3}{10}-1.26+(-2.3)=0.3-1.26-2.3$
$=-3.26$

**Cumulative Review**

**61.** $(231)(14)=3234$

**63.** $(19)(-15)=-285$

**8.3 Exercises**

**1.** If one factor has 3 decimal places and the second factor has 2 decimal places, the product has <u>5</u> decimal places.

**3.** When we divide $4.62\overline{)12.7}$, we rewrite the equivalent division problem $\underline{462\overline{)1270}}$ and then divide.

**5.** $0.03\times0.07\to7\times3=21$
$0.03\times0.07=0.0021$

**7.** $0.05\times0.07\to5\times7=35$
$0.05\times0.07=0.0035$

**9.**
$$\begin{array}{r} 7.43 \\ \underline{\times 8.3} \\ 2229 \\ \underline{5944\phantom{0}} \\ 61.669 \end{array}$$

**11.**
$$\begin{array}{r} 15.2 \\ \underline{\times 3.1} \\ 152 \\ \underline{456\phantom{0}} \\ 47.12 \end{array}$$

**13.** $(-5)(1.41)=-7.05$

**15.** $(-4.23)(2.7)=-11.421$

**17.** $(-25)(-0.613)=15.325$

**19.** $(12.1)(-2.81)=-34.001$

**21.** $0.1498\times100=14.98$

**23.** $85.54\times10{,}000=855{,}400$

**25.** $24\times10^4 = 240,000$

**27.** $0.2\times10^4 = 2000$

**29.** $\begin{array}{r} 2.16 \\ 8\overline{)17.28} \\ \underline{16} \\ 12 \\ \underline{8} \\ 48 \\ \underline{48} \end{array}$

**31.** $\begin{array}{r} 0.62 \\ 12\overline{)7.44} \\ \underline{72} \\ 24 \\ \underline{24} \end{array}$

**33.** $\begin{array}{r} 0.0565 \\ 64\overline{)3.6160} \\ \underline{320} \\ 416 \\ \underline{384} \\ 320 \\ \underline{320} \end{array}$

**35.** $\begin{array}{r} 3.451 \\ 24\overline{)84.824} \\ \underline{72} \\ 108 \\ \underline{96} \\ 122 \\ \underline{120} \\ 24 \\ \underline{24} \end{array}$

**37.** $\begin{array}{r} 0.23 \\ 14\overline{)3.25} \\ \underline{28} \\ 45 \\ \underline{42} \\ 3 \end{array}$

**39.** $3.6\overline{)-0.2988} \to \begin{array}{r} -0.08 \\ 36\overline{)-2.988} \\ \underline{288} \\ 10 \end{array}$

**41.** $-1.7\overline{)-20.8} \to \begin{array}{r} 12.23 \\ 17\overline{)208.00} \\ \underline{17} \\ 38 \\ \underline{34} \\ 40 \\ \underline{34} \\ 60 \\ \underline{51} \\ 9 \end{array} \to 12.24$

**43.** $0.27\overline{)8.343} \to \begin{array}{r} 30.9 \\ 27\overline{)834.3} \\ \underline{81} \\ 243 \\ \underline{243} \\ 60 \\ \underline{54} \\ 6 \end{array}$

**45.** $13.7592 \div 5.88 \rightarrow 588\overline{)1375.92}$ (quotient 2.34)

$$\begin{array}{r} 2.34 \\ 588\overline{)1375.92} \\ \underline{1176} \\ 1999 \\ \underline{1764} \\ 2352 \\ \underline{2352} \end{array}$$

**47. 45.** $1.8\overline{)5} \rightarrow 18\overline{)50.0} \rightarrow 2.\overline{7}$

$$\begin{array}{r} 2.7 \\ 18\overline{)50.0} \\ \underline{36} \\ 140 \\ \underline{126} \\ 14 \end{array}$$

**49.** $1.1\overline{)0.7} \rightarrow 11\overline{)7.0} \rightarrow 0.\overline{63}$

$$\begin{array}{r} 0.63 \\ 11\overline{)7.0} \\ \underline{66} \\ 40 \\ \underline{33} \\ 7 \end{array}$$

**51.** $2.2\overline{)11.3} \rightarrow 22\overline{)113.000} \rightarrow 5.1\overline{36}$

$$\begin{array}{r} 5.136 \\ 22\overline{)113.000} \\ \underline{110} \\ 30 \\ \underline{22} \\ 80 \\ \underline{66} \\ 140 \\ \underline{132} \\ 8 \end{array}$$

**53.** $6.6\overline{)200} \rightarrow 66\overline{)2000.00} \rightarrow 3.\overline{30}$

$$\begin{array}{r} 30.30 \\ 66\overline{)2000.00} \\ \underline{198} \\ 200 \\ \underline{198} \\ 2 \end{array}$$

**55.** $\frac{11}{6} = 1.8\overline{3} = 1.83$, nearest hundredth

**57.** $12\frac{2}{15} = 12.1\overline{3} = 12.13$, nearest hundredth

**59.** $\frac{1}{3} = 0.\overline{3}$

**61.** $\frac{7}{9} = 0.\overline{7}$

**63.**

$20.35 \div 0.44 \rightarrow 44\overline{)2035}$

$$\begin{array}{r} 46.25 \\ 44\overline{)2035} \\ \underline{176} \\ 275 \\ \underline{264} \\ 110 \\ \underline{88} \\ 220 \\ \underline{220} \end{array}$$

**65.** $9\overline{)2.0} \rightarrow \frac{2}{9} = 0.\overline{2}$

$$\begin{array}{r} 0.2 \\ 9\overline{)2.0} \\ \underline{18} \\ 2 \end{array}$$

**67.** $-3.5 \times 4.24 = -14.84$

**69.** $0.7 \times 0.8 = 0.56$

**71.** (a) $\frac{1}{300} = 0.00\overline{3}$ (b) $\frac{1}{30} = 0.0\overline{3}$

**73.** $-562.53 \div 13.123$
$= -42.8659605273\cdots$
$= -42.866$, nearest thousandth

**75.** $\frac{4}{9} = 0.44\cdots$, $\frac{5}{9} = 0.55\cdots$

## Cumulative Review

**77.** $12x = 96 \Rightarrow x = \frac{96}{12} = 8$

check: $12(8) \stackrel{?}{=} 96$, $96 = 96$

**79.** $x + 45 = 17 \Rightarrow x = 17 - 45 = -28$

check: $-28 + 45 \stackrel{?}{=} 17$, $17 = 17$

## 8.4 Exercises

**1.** $x + 3.7 = 9.8 \Rightarrow x = 9.8 - 3.7 = 6.1$

**3.** $y - 2.5 = 6.95 \Rightarrow y = 6.95 + 2.5 = 9.45$

**5.** $2.9 + x = 5 \Rightarrow x = 5 - 2.9 = 2.1$

**7.** $x + 2.5 = -9.6 \Rightarrow x = -9.6 - 2.5 = -12.1$

**9.** $2x = 11.24 \Rightarrow x = \frac{11.24}{2} = 5.62$

**11.** $5.1x = 25.5 \Rightarrow x = \frac{25.5}{5.1} = 5$

**13.** $-5.6x = -19.04 \Rightarrow x = \frac{-19.04}{-5.6} = 3.4$

**15.** $-5.2x = 26 \Rightarrow x = \frac{26}{-5.2} = -5$

**17.** $-3x - 5.3 = 11.23 \Rightarrow -3x = 16.53$

$x = \frac{16.53}{-3} = -5.51$

**19.** $0.9x + 8.7 = 15.9 \Rightarrow 0.9x = 7.2$

$x = \frac{7.2}{0.9} = 8$

**21.** $2(x - 4) = 26.4 \Rightarrow 2x - 8 = 26.4$

$2x = 34.4 \Rightarrow x = \frac{34.4}{2} = 17.2$

**23.** $2(x + 3.2) = x + 9.9 \Rightarrow 2x + 6.4 = x + 9.9$
$x = 9.9 - 6.4 = 3.5$

**25.** $0.3x + 0.2 = 1.7 \Rightarrow 3x + 2 = 17$

$3x = 15 \Rightarrow x = \frac{15}{3} = 5$

**27.** $0.8x + 0.6 = 5.4 \Rightarrow 8x + 6 = 54$

$8x = 48 \Rightarrow x = \frac{48}{8} = 6$

**29.** $0.12x + 1.1 = 1.22 \Rightarrow 12x + 110 = 122$
$12x = 12 \Rightarrow x = 1$

**31.** $0.15x + 0.23 = 1.43 \Rightarrow 15x + 23 = 143$
$15x = 120 \Rightarrow x = 8$

**33.** $5.6 + x = -4.8 \Rightarrow x = -4.8 - 5.6 = -10.4$

**35.** $3.4x = 10.2 \Rightarrow x = \frac{10.2}{3.4} = 3$

**37.** $3(x + 1.4) = 6.9 \Rightarrow 3x + 4.2 = 6.9$

$3x = 2.7 \Rightarrow x = \frac{2.7}{3} = 0.9$

**39.** $6x+10.5=x+21 \Rightarrow 5x=10.5$
$x=\frac{10.5}{5}=2.1$

**41.** (a) $\frac{901 \text{ mi}}{34 \text{ gal}}=26.5 \text{ mi/gal}$
(b) $\frac{1484 \text{ mi}}{26.5 \text{ mi/gal}}=56 \text{ gal}$

**43.** (a) $17(0.24)=\$4.08$
(b) $17(0.14)=\$2.38$
(c) $4.08-2.38=\$1.70$

**45.** (a) $\frac{65 \text{ liters}}{\frac{2.5 \text{ liters}}{\text{container}}}=26 \text{ containers}$
(b) $\frac{\$5.70}{\text{liter}}\cdot\frac{2.5 \text{ liters}}{\text{container}}=\frac{\$14.25}{\text{container}}$

**47.** $2(24.95)+3(12.98)+54.25=\$143.09$

**49.**

| total cost per day | × | number of days | + | charge/mi over 200 mi | × | number of mi over 200 | = | total cost |
|---|---|---|---|---|---|---|---|---|
| ↓ | | ↓ | | ↓ | | ↓ | | ↓ |
| 18.95 | × | 3 | | +0.12 | | ×(423 − 200) | | = \$83.61 |

**51.** $4.95+0.07(75)+0.10(20)=12.2$
Single rate plan: \$12.20
$0.10(75)+0.15(20)=10.5$
No Fee Plan: \$10.50 is the better deal.

**53.** (a) $Q=$ number of quarters
$5Q=$ number of dimes
(b) $0.25Q+0.10(5Q)=4.50$
(c) $0.75Q=4.50 \Rightarrow Q=\frac{4.50}{0.75}=6$
$Q=6$ quarters, $5Q=30$ dimes

**53.** (d)
$0.25(6)+0.10(5\cdot 6)\stackrel{?}{=}4.50, \ 4.50=4.50$

**55.** $0.05n+0.10(5n)=2.20 \Rightarrow 0.55n=2.20$
$n=4$ nickels, $5n=20$ dimes

**57.** $23.098x=103.941 \Rightarrow x=\frac{23.098}{103.941}=4.5$

**59.** $x+3.0012=21.566$
$x=21.566-3.0012=18.5648$

**61.** $a=\frac{v}{t} \Rightarrow v=at=15(3.5)=52.5 \text{ ft/sec}$

### Cumulative Review

**63.** $\frac{4}{5}=0.8$

**65.** $\frac{1}{3}=0.\bar{3}$

**67.** $25-12\frac{1}{3}=24\frac{3}{3}-12\frac{1}{3}=12\frac{2}{3}$ lb

### How Am I Doing? Sections 8.1-8.4

**1.** $\frac{2}{100}=0.02$

**2.** $0.027=\frac{27}{1000}$

**3.** $0.56<0.566$

**4.** $4212.65133=4212.651$
nearest thousandth

**5.** $35+4.73+0.623=40.353$

**6.** $-81.14-15.313=-96.453$

**7.** $2.3x+3.1y+4.4x=6.7x+3.1y$

**8.** $y-0.921=5.8-0.921=4.879$

**9.** $(-3.23)(1.61) = -5.2003$

**10.** $0.2783 \times 10^3 = 278.3$

**11.** $13.806 \div 2.6 = 5.31$

**12.** $\frac{23}{6} = 3.8333\cdots = 3.8\overline{3}$

**13.** $-2.1x - 4.4 = 3.16 \Rightarrow -2.1x = 7.56$
$x = -3.6$

**14.** $2(x + 4.5) = x + 9.8 \Rightarrow 2x + 9 = x + 9.8$
$x = 0.8$

**15.** $40(\$9.25) + (53 - 40)(\$13.88) = \$550.44$

**8.5 Exercises**

**1.** To estimate 10% of a number, we can delete the last digit.

**3.** We can find 15% by adding 5% and 10%.

**5.** We can find 6% by adding 5% and 1%.

**7.** 10% of 801: 80

**9.** $5\%(801) \approx \frac{1}{2}(80) = 40$

**11.** $20\%(801) \approx 2(80) = 160$

**13.** $1\%(205) \approx 2$

**15.** $7\%(205) = 5\%(205) + 2\%(205)$
$\approx \frac{1}{2}(20) + 2(2) = 14$

**17.** $30\%(205) = 3(10\%(205)) \approx 3(20) = 60$

**19.** $10\%(1007) \approx 100$

**21.** $15\%(1007) = 10\%(1007) + 5\%(1007)$
$= 10\%(1007) + \frac{1}{2} \cdot 10\%(1007)$
$\approx 100 + \frac{1}{2}(100) = 150$

**23.** $20\%(1007) = 2(10\%(1007))$
$\approx 2(100) = 200$

**25.** $1\%(3015) \approx 30$

**27.** $2\%(3015) = 2(1\%(3015)) \approx 2(30) = 60$

**29.** $3\%(3015) = 3(1\%(3015) \approx 3(30) = 90$

**31.** $10\%(320,050) \approx 10\%(320,000)$
$= 32,000$

**33.** $5\%(320,050) \approx 5\%(320,000)$
$= \frac{1}{2} \cdot 10\%(320,000) = \frac{1}{2} \cdot 32,000$
$= 16,000$

**35.** $15\%(320,050) \approx 15\%(320,000)$
$= 10\%(320,000) + 5\%(320,000)$
$= 32,000 + \frac{1}{2}(10\%(320,000))$
$= 32,000 + \frac{1}{2}(32,000) = 48,000$

**37.** $1\%(250,030) \approx 1\%(250,000) = 2500$

**39.** $4\%(250,030) \approx 4(1\%(250,000))$
$= 4(2500) = 10,000$

**41.** $8\%(250,030) \approx 8(1\%(250,000))$
$= 8(2500) = 20,000$

**43.** $5\%(\$19,999) \approx 5(1\%(\$20,000))$
$= 5(\$200) = \$1000$

**45.** $6\%(\$139,999) \approx 6(1\%(\$140,000))$
$= 6(\$1400) = \$8400$

**47.** (a) $20 + 200 + 2(50) = \$320$
(b) $40\%(320) \approx 4(10\%(300))$
$= 4(30) = \$120$
(c) $320 - 120 = \$200$

**49.** $0.60(19.99 + 199 + 2(49.99)) = \$191.38$
$200 - 191.38 = \$8.62$

**Cumulative Review**

**51.** $\frac{5}{n} = \frac{7.5}{18} \Rightarrow 7.5n = 90 \Rightarrow n = 12$

**53.** $1 - \frac{1}{4} - \frac{3}{8} = \frac{8-2-3}{8} = \frac{3}{8}$ of the tank

**Putting Your Skills to Work, Problems for Individual Analysis**

**1.** $10\%(\$240,000) = \$24,000$
$5\%(\$240,000) = \frac{1}{2} \cdot 10\%(\$240,000)$
$= \frac{1}{2} \cdot \$24,000 = \$12,000$
$1\%(\$240,000) = \$2400$

**2.** (a) $5\% + 1\% \rightarrow 12,000 + 2400 = \$14,400$
(b) $2\% = 2(1\%) \rightarrow 2(2400) = \$4800$
(c) $\$4800 + \$14,400 = \$19,200$

**3.** $\$240,000 - \$19,200 = \$220,800$

**Putting Your Skills to Work, Problems for Group Investigation and Study**

**1.** (a) $\$1800 = 1\% \cdot x \Rightarrow x = \$180,000$
(b) $\$240,000 - \$180,000 = \$60,000$
(c) $2\%(\$240,000) = 2(1\%(\$240,000))$
$= 2(\$2400) = \$4800$
(d) $1\frac{1}{2}\%(\$180,000)$
$= 1\%(\$180,000) + \frac{1}{2} \cdot 1\%(\$180,000)$
$= \$1800 + \frac{1}{2} \cdot \$1800 = \$2700$
(e) $\$60,000 + \$4800 + \$2700 = \$67,500$

| Sale Price | $240,000 |
|---|---|
| (a) Loan amount | $180,000 |
| (b) Down payment | $60,000 |
| (c) Escrow fee | $4800 |
| (d) Loan fees | $2700 |
| (e) Total cash needed | $67,500 |

**2.** (a) $\$160,000 - \$90,000 = \$70,000$
(b) $6\%(\$160,000) = \$9600$
(c) $2\%(\$160,000) = \$3200$
(d) $\$70,000 - \$9600 - \$3200 = \$57,200$
(e) Toward the purchase they have $\$57,200 + \$10,000 = \$67,200$. They need $67,500 so they are short $300.

| Sale Price of Townhouse | $160,000 |
|---|---|
| Payoff of existing loan | $90,000 |
| (a) Funds left after payoff | $70,000 |
| (b) Real estate fee | $9600 |
| (c) Escrow fees | $3200 |
| (d) Funds left for purchase | $57,200 |

**8.6 Exercises**

**1.** To change a percent to a decimal, move the decimal point 2 places to the left and drop the % sign.

**3.** $\frac{31}{100} = 31\%$

**5.** $\frac{63}{100} = 63\%$

**7.** $\frac{16}{100} = 16\%$

**9.** $\frac{113}{100} = 113\%$

**11.** $\frac{0.9}{100} = 0.9\%$

**13.**

| Decimal | Percent |
|---|---|
| 0.576 | 57.6% |
| 0.249 | 24.9% |
| 0.003 | 0.3% |
| 1.546 | 154.6% |

**15.**

| Decimal | Percent |
|---|---|
| 3.7 | 370.0% |
| 0.238 | 23.8% |
| 0.006 | 0.6% |
| 12.882 | 1288.2% |

**17.** $78\% = \frac{78}{100} = 0.78$

**19.** $53.8\% = 0.538$

**21.** $0.0024 = 0.24\%$

**23.** $2.33\% = 0.0233$

**25.** $0.03413 = 3.413\%$

**27.**

| Fraction Form | Decimal Form | Percent Form |
|---|---|---|
| $\frac{4}{5}$ | 0.8 | 80% |
| $\frac{27}{100}$ | 0.27 | 27% |
| $\frac{7}{1000}$ | 0.007 | 0.7% |
| $4\frac{1}{3}$ | $4.\overline{3}$ | $433.\overline{3}\%$ |

**29.**

| Fraction Form | Decimal Form | Percent Form |
|---|---|---|
| $\frac{5}{16}$ | 0.3125 | 31.25% |
| $2\frac{3}{5}$ | 2.6 | 260% |
| $\frac{1}{400}$ | 0.0025 | $\frac{1}{4}\%$ |
| $6\frac{1}{2}$ | 6.5 | 650% |

**31.** $\frac{35}{40} = 0.875 = 87.5\%$

**33.** $\frac{3}{4}\% = \frac{3}{4} \cdot \frac{1}{100} = \frac{3}{400}$

**35.** (a) $\frac{14}{40} = 0.35 = 35\%$

(b) $22.3\% = \frac{22.3}{100} = \frac{223}{1000}$

**37.** $1\frac{1}{4}\% = \frac{5}{4} \cdot \frac{1}{100} = \frac{1}{80}$ (5 and 100 cancelled; 100 → 20)

**39.** $\frac{1}{40} = 0.025 = 2.5\%$

**41.** $\frac{5}{7} = 0.714285 = 71.4285\% = 71.43\%$

**43.** $\frac{5}{6} = 0.8333\cdots = 83.333\cdots\% = 83.33\%$

**45.** $5.5\% = \frac{5.5}{100} = \frac{55}{1000} = \frac{11}{200}$

**47.** $11.5\% = \frac{11.5}{100} = \frac{115}{1000} = \frac{23}{200}$

**49.** $15.9\% = \frac{15.9}{100} = \frac{159}{1000}$

**51.**

| Fraction | $\frac{3}{5}$ | $\frac{4}{5}$ |
|---|---|---|
| Percent | 60% | 80% |

**53.**

| Fraction | $4\frac{2}{8}$ | $4\frac{3}{8}$ |
|---|---|---|
| Percent | 425% | 437.5% |

**Cumulative Review**

**55.** $2x = 330 \Rightarrow x = \frac{330}{2} = 165$

**57.** $x = \frac{1}{3} \cdot 69 = 23$

**59.** $5.1 - 3.7 = 1.4$ million mi$^2$

**8.7 Exercises**

**1.** 35 is much more than $\frac{1}{2}$ of 40.

**3.** 9 is very close to 10 so it cannot be $\frac{1}{4}$ of 10.

**5.** $x = 0.32 \times 84 = 26.88$

**7.** $n = 0.24 \times 145 = 34.8$

**9.** $n = 0.46 \times 60 = 27.6$

**11.** $n = 150\% \times 40 = 1.5 \times 40 = 60$

**13.** $0.15(12.95) = \$1.94$

**15.** $0.60(650) = 390$ students

**17.** $60 = n \times 30 \Rightarrow n = \frac{60}{30} = 2 = 200\%$

**19.** $70 = p \times 650 \Rightarrow p = \frac{70}{650} = 0.10769\cdots$

$p = 10.77\%$

**21.** $n \times 90 = 18,\ n = \frac{18}{90} = 0.2 = 20\%$

**23.** $\frac{15}{80} = 0.1875 = 18.75\%$

**25.** $\frac{2}{5} = 0.4 = 40\%$

**27.** $56 = 0.70 \times n \Rightarrow n = 80$

**29.** $24 = 0.40 \times n \Rightarrow n = 60$

**31.** $80 = 0.20 \times n,\ n = \dfrac{80}{0.20} = 400$

**33.** $10 = 0.05 \times n \Rightarrow n = 200$

**35.** $8 = 1.25 \times d \Rightarrow d = 6.4$ mi

**37.** $x + 0.6x = 128$

$$1.6x = 128$$

$$x = \frac{128}{1.6} = \$80$$

**39.** $x + 0.9x = 95 \Rightarrow 1.9x = 95 \Rightarrow x = \$50$

**41.** $44 = 0.50 \times n \Rightarrow n = 88$

**43.** $1.25 \times 85 = n \Rightarrow n = 106.25$

**45.** $n = 0.27 \times 78 \Rightarrow n = 21.06$

**47.** $53.94 = n \times 87 \Rightarrow n = 0.62 = 62\%$

**49.** $135 = 0.45 \times n,\ n = \dfrac{135}{0.45} = 300$

**51.** $110 = n \times 440$

$$n = \frac{110}{440} = 0.25 = 25\%$$

**53.** $0.73 \times 130$ million $= 94.9$ million

**55.** $\dfrac{86}{92} = 0.93478\cdots = 93.478\cdots\% \to 93\%$

**57.** $34.5\% - 24.5\% = 10\%$

**59.** $100\% - (13 + 10 + 27 + 33 + 10)\% = 7\%$

**61.** $0.13(\$1250) = \$162.50$

**63.** $n = 0.673(348.9) = 234.8097$

**65.** $368 = 0.20n \Rightarrow n = 1840$

**67.** $0.346(1,400,000) = n = 484,400$

**Cumulative Review**

**69.** $2x + 3 = 13 \Rightarrow 2x = 10 \Rightarrow x = 5$

**71.** $5x - 3 = 3x + 9 \Rightarrow 2x = 12 \Rightarrow x = 6$

**8.8 Exercises**

**1.** Since 100% of 80 is 80, it is obvious that 150% of 80 is greater than 80.

**3.** $p = 22,\ b = 250,\ a = 55$

**5.** $p = 95,\ b = 380,\ a = a$

**7.** $p = 69,\ b = b,\ a = 8230$

**9.** $p = p,\ b = 90,\ a = 70$

**11.** $p = p,\ b = 25,\ a = 10$

**13.** $p = 160,\ b = b,\ a = 400$

**15.** $\dfrac{a}{300} = \dfrac{24}{100} \Rightarrow a = 72$

**17.** $\dfrac{a}{30} = \dfrac{250}{100} \Rightarrow a = 75$

**19.** $\dfrac{a}{4000} = \dfrac{0.6}{100} \Rightarrow a = 24$

**21.** $\dfrac{82}{b} = \dfrac{50}{100} \Rightarrow b = 164$

**23.** $\frac{90}{b}=\frac{150}{100}\Rightarrow b=60$

**25.** $\frac{4000}{b}=\frac{0.8}{100}\Rightarrow b=500{,}000$

**27.** $\frac{70}{280}=\frac{p}{100}\Rightarrow p=25$

**29.** $\frac{3.5}{140}=\frac{p}{100}\Rightarrow p=2.5$

**31.** $\frac{90}{5000}=\frac{p}{100}\Rightarrow p=1.8$

**33.** $\frac{a}{350}=\frac{26}{100}\Rightarrow a=91$

**35.** $\frac{540}{b}=\frac{180}{100}\Rightarrow b=300$

**37.** $\frac{75}{400}=\frac{p}{100}\Rightarrow p=18.75$; 18.75%

**39.** $\frac{a}{650}=\frac{0.2}{100}\Rightarrow a=1.3$

**41.** $\frac{15.2}{25}=\frac{p}{100}\Rightarrow p=60.8$; 60.8%

**43.** $\frac{68}{b}=\frac{40}{100}\Rightarrow b=170$

**45.** $\frac{94.6}{220}=\frac{p}{100}\Rightarrow p=43$; 43%

**47.** $\frac{a}{380}=\frac{12.5}{100}\Rightarrow a=47.5$

**49.** $\frac{a}{5600}=\frac{0.05}{100}\Rightarrow a=2.8$

**51.** $\frac{a}{250}=\frac{80}{100}\Rightarrow a=200$ apts

**53.** (a) $100\%-60\%=40\%$

(b) $\frac{a}{150}=\frac{40}{100}\Rightarrow a=60$ openings

**55.** (a) $\frac{42.9}{1950}=\frac{p}{100}\Rightarrow p=2.2$; 2.2%

(b) $100\%-2.2\%=97.8\%$

**57.** $\frac{90}{b}=\frac{75}{100}\Rightarrow b=120$ units

**59.** $\frac{6}{9}=\frac{p}{100}\Rightarrow p=66.\overline{6}$; 67%

**61.** $0.1925(798)=153.615\rightarrow 153.62$

**63.** $0.18(0.20(\$3300))=\$118.80$

**Cumulative Review**

**65.** $A=LW=7(4)=28\text{ in.}^2$

**67.** $A=s^2=2^2=4\text{ ft}^2$

**8.9 Exercises**

**1.** Principal: the amount deposited or borrowed

**3.** Time: the period of time interest is calculated

**5.** Commission = commission rate × total sales

**7.** $0.15(2350) = \$352.50$

**9.** $0.25(13,500) = \$3375$

**11.** $\frac{379.5}{1725} = 0.22 = 22\%$

**13.** $n \times 1800 = 360$
$n = 0.2 = 20\%$

**15.** $4500 - 0.1(4500) = \$4050$

**17.** $1.08(35,500) = \$38,340$

**19.** $1.15(760) = 874$ people

**21.** (a) $0.07(3000) = \$210$
(b) $3000 + 210 = \$3210$

**23.** $I = PRT = 500(0.08)(0.5) = \$20$

**25.** (a) $0.20(4150) = \$830$
(b) $1500 + 830 = \$2330$

**27.** $\frac{3500 + 3500(0.08)(1)}{12} = \$315$

**29.** $265 - 0.15(265) = 225.25$ lb

**31.** (a) $2(289) + 421 = \$999$
(b) $0.7(999) = \$699.30$

**33.** $0.05x = 35 \Rightarrow x = \$700$

**35.** (a) $1.022x = 1232 \Rightarrow x = \$1205.48$
(b) $1.022(1232) = \$1259.10$

**37.** (a) $0.03(35)(0.11)(320) = \$36.96$
(b) $0.07(35)(0.11)(320) - 50 = \$36.24$

**Cumulative Review**

**39.** $A = bh = 6(4) = 24 \text{ cm}^2$

**Chapter 8 Review**

**1.** 6.23: Six and twenty-three hundredths

**2.** 0.679: Six hundred seventy-nine thousandths

**3.** 7.0083: Seven and eighty-three ten thousandths

**4.** Forty-six and 85/100

**5.** $4.267 = 4\frac{267}{1000}$

**6.** $43.91 = 43\frac{91}{100}$

**7.** $32\frac{761}{1000} = 32.761$

**8.** $54\frac{26}{1000} = 54.026$

**9.** $0.523 < 0.524$

**10.** $0.16 < 0.168$

**11.** $842.8569 = 842.86$, nearest hundredth

**12.** $406.7809 = 406.781$, nearest thousandth

**13.** $0.52 + 8.11 = 8.63$

**14.** $-5.2 + 0.236 = -4.964$

**15.** $0.588 + 36 + 8.43 = 45.018$

**16.** $25.98 - 2.33 = 23.65$

**17.** $-2.12 - 9.67 = -11.79$

**18.** $-9.355 - 2.48 = -11.835$

**19.** $(-9.2) + (-5.4) = -14.6$

**20.** $4.32 - (-6.43) = 10.75$

**21.** $-7 - (-6.67) = -0.33$

**22.** To subtract numbers in decimal notation, we line up the decimal points.

**23.** When adding $85 + 36.5$, we rewrite 85 as 85.0 so that we can line up the decimal points.

**24.** $x - 9.3\big|_{x=0.6} = 0.6 - 9.3 = -8.7$

**25.** $y + 17.2\big|_{y=-2.3} = -2.3 + 17.2 = 14.9$

**26.** $4.6x + 7.2x = 11.8x$

**27.** $8.6x + 3.9x = 12.5x$

**28.** $23 + 25 + 21 + 24 = \$93$

**29.** $100 - 54.56 - 21.06 = \$24.38$

**30.** $0.091 \times 0.06 = 0.00546$

**31.** $0.082 \times 0.02 = 0.00164$

**32.** $5.68 \times 7.21 = 40.9582$

**33.** $2.62 \times 7.33 = 19.2046$

**34.** $(3.01)(-41.25) = -124.1625$

**35.** $(-5.6)(-9.01) = 50.456$

**36.** If one factor has 2 decimal places and the second factor has 3 decimal places, the product has five decimal places.

**37.** If one factor has 3 decimal places and the second factor has 4 decimal places, the product has seven decimal places.

**38.** $0.1249 \times 100 = 12.49$

**39.** $3.24 \times 1000 = 3240$

**40.** $41 \times 10^5 = 4,100,000$

**41.** $40(8.30) + 9(12.45) = \$444.05$

**42.** $4500 + 0.12(44,322.6 - 36,000)$
$= \$5498.71$

**43.** $8.66 \div 12 = 0.72$

**44.** $-3.25 \div 5.1 = -0.64$

**45.** $-8.52 \div -7.2 = 1.18$

**46.** $16.221 \div 0.33 = 49.1\overline{54}$

**47.** $13.01 \div 0.33 = 39.\overline{42}$

**48.** When we divide $7.21\overline{)25.9}$, we rewrite the equivalent division problem $721\overline{)2590}$ and then divide.

**49.** $12\overline{)4.349}$

**50.** $\frac{86}{8} = 10.75$

**51.** $9\frac{1}{5} = 9.2$

**52.** $\frac{13}{9} = 1.\overline{4} = 1.44$, nearest hundredth

**53.** (a) $38.5 \div 3.5 = 11$ containers
(b) $3.5(6.20) = \$21.70$

**54.** $x - 2.68 = 8.23 \Rightarrow x = 8.23 + 2.68 = 10.91$

**55.** $-1.6x = 3.68 \Rightarrow x = \frac{3.68}{-1.6} = -2.3$

**56.** $2x + 2.4 = 8.7 \Rightarrow x = \frac{8.7 - 2.4}{2} = 3.15$

**57.** $3x - 2.8 = 2x + 4.2 \Rightarrow x = 4.2 + 2.8 = 7$

**58.** $0.05(5d) + 0.10d = 14 \Rightarrow 0.35d = 14$
$d = \frac{14}{0.35} = 40$ dimes
$5d = 200$ nickels

**59.** $10\%(176,001) \approx 10\%(176,000) = 17,600$

**60.** $1\%(176,001) \approx 1\%(176,000) = 1760$

**61.** $7\%(176,001) \approx 7\%(176,000)$
$= \frac{1}{2} \cdot 10\%(176,000) + 2 \cdot 1\%(176,000)$
$= 12,320$

**62.** $20\%(176,001) \approx 2 \cdot 10\%(176,000)$
$= 35,200$

**63.** $\frac{85}{100} = 85\%$

**64.** $\frac{132}{110} = 1.2 = 120\%$

**65.** $5.7\% = 0.057$

**66.** $0.016 = 1.6\%$

**67.** $124\% = 1.24$

**68.**

| Decimal Form | Percent Form |
|---|---|
| 0.379 | 37.9% |
| 0.428 | 42.8% |
| 0.005 | 0.5% |
| 3.47 | 347% |

**69.**

| Decimal Form | Percent Form |
|---|---|
| 1.2 | 120% |
| 0.035 | 3.5% |
| 0.0025 | 0.25% |
| 0.567 | 56.7% |

**70.** $\frac{56}{168} = 0.\overline{3} = 33.\overline{3}\% = 33\frac{1}{3}\%$

**71.** $81.3 = 81\frac{3}{10}$

**72.**

| Fraction Form | Decimal Form | Percent Form |
|---|---|---|
| $\frac{3}{8}$ | 0.375 | 37.5% |
| $\frac{18}{25}$ | 0.72 | 72% |
| $\frac{1}{200}$ | 0.005 | $0.5\% = \frac{1}{2}\%$ |
| $7\frac{2}{5}$ | 7.4 | 740% |

**73.**

| Fraction Form | Decimal Form | Percent Form |
|---|---|---|
| $\frac{3}{4}$ | 0.75 | 75% |
| $\frac{14}{25}$ | 0.56 | 56% |
| $\frac{1}{500}$ | 0.002 | $0.2\% = \frac{1}{5}\%$ |
| $3\frac{6}{15}$ | 3.4 | 340% |

**74.** $82 = 0.25n \Rightarrow n = 328$

**75.** $0.30(90) = 27$

**76.** $15 = p \cdot 300 \Rightarrow p = 0.05 = 5\%$

**77.** $0.45(120) = 54$

**78.** $\frac{18}{1.15} \approx \$15.65$

**79.** $26 = 0.10n \Rightarrow n = 260$

**80.** $0.07(17,000) = \$1190$

**81.** $\frac{4.25}{32.40} = 0.1311772839506\cdots \approx 13.12\%$

**82.** $p = 85,\ b = 400,\ a = a$

**83.** $p = p,\ b = 100,\ a = 20$

**84.** $9 = 0.45b \Rightarrow b = 20$

**85.** $22 = p(88) \Rightarrow p = 0.25 = 25\%$

**86.** (a) The original price of the dishwasher in the advertisement is $\underline{225 + 75 = \$300}$.

(b) $75 = p(300) \Rightarrow p = 0.25 = 25\%$
The dishwasher is marked down $\underline{25\%}$.

**87.** (a) $100\% - 70\% = 30\%$
(b) $0.30(250) = 75$

**88.** $0.18(13,250) = \$2385$

**89.** $\frac{59,500 - 50,0000}{50,0000} = 0.19 = 19\%$

**90.** $I = PRT = 7500(0.13)(1) = \$975$

**How Am I Doing? Chapter 8 Test**

**1.** 207.402: Two hundred seven and four hundred two thousandths

**2.** $0.013 = \frac{13}{1000}$

**3.** $\frac{51}{100} = 0.51$

**4.** $0.45 > 0.412$

**5.** $746.136 = 746.14$, nearest hundredth

**6.** $12.93 + 0.21 = 13.14$

**7.** $x + 0.12\big|_{x=-2.07} = -2.07 + 0.12 = -1.95$

**8.** $3.1x + 2.01y + 1.06x = 4.16x + 2.01y$

**9.** $18.81 - 6.17 = 12.64$

**10.** $(-13.2) - (-7.1) = -6.1$

**11.** $(8.24)(1.2) = 9.888$

**12.** $(4.72)(10^3) = 4720$

**13.** $15.75 \div 3.5 = 4.5$

**14.** $\frac{19}{3} = 6.\overline{3}$

**15.** $\frac{7}{100} = 7\%$

**16.** (a) $\frac{76}{95} = 0.8$

(b) $\frac{76}{95} = 0.8 = 80\%$

**17.** (a) $0.10 = \frac{10}{100} = \frac{1}{10}$

(b) $0.10 = \frac{10}{100} = 10\%$

**18.** (a) $5\% = \frac{5}{100} = \frac{1}{20}$

(b) $5\% = \frac{5}{100} = 0.05$

**19.** $0.24(3300) = \$792$

**20.** $\frac{249}{1.2} = \$207.50$

**21.** $5500 + 0.12(34{,}100.5 - 30{,}000)$

$= \$5992.06$

**22.** (a) $0.134

(b) $1.604 - 1.382 = \$0.222$

**23.** $\% \text{ increase} = \frac{1.457 - (1.457 - 0.065)}{1.457 - 0.065}$

$\% \text{ increase} = \frac{1.457 - 1.392}{1.392} = 0.0466\cdots$

$\% \text{ increase} = 4.7\%$

**24.** $0.5x + 0.2x = 2.8 \Rightarrow x = \frac{2.8}{0.7} = 4$

**25.** $2(y + 1.3) = 7.8 \Rightarrow y = \frac{7.8}{2} - 1.3 = 2.6$

**26.** (a) $\frac{577.2}{33.1} \approx 17.4$ mi/gal

(b) $33.1(1.27) = \$42.04$

**27.** $12 = 0.30n \Rightarrow n = \frac{12}{0.30} = 40$

**28.** $5 = p(250) \Rightarrow p = \frac{5}{250} = 0.02 = 2\%$

**29.** $a = 0.20(48) = 9.6$

**30.** $0.15(10{,}500) = \$1575$

**31.** $I = PRT = 8200(0.11)(2) = \$1804$

**32.** One Step: $0.80(850) - 50 = \$630$

One Touch:

$0.85(850) - 100 = \$622.50$, better buy

**Cumulative Test for Chapters 1-8**

**1.** $x+2x+3xy=3x+3xy$

**2.** $5x-x+2x=4x+2x=6x$

**3.** $3-6x+2xy-6-4xy=-3-6x-2xy$

**4.** $P=4x=4(8)=32$ ft

**5.** $A=LW=11(9)=99\text{ ft}^2$

**6.** $-10,000+25,000=\$15,000$ profit

**7.** $4x-2\big|_{x=-1}=4(-1)-2=-6$

**8.** $4x-2\big|_{x=10}=4(10)-2=38$

**9.** $4x-2\big|_{x=\frac{1}{2}}=4\left(\frac{1}{2}\right)-2=0$

**10.** $\frac{250}{30}=8\frac{1}{3}$ cal/min

**11.** $\frac{18y}{27y^2}=\frac{2}{3y}$

**12.** $(x^3)^2=x^{3\cdot 2}=x^6$

**13.** $(4xy)^3=4^3x^3y^3=64x^3y^3$

**14.** $\frac{2}{3}=\frac{26}{x}\Rightarrow 2x=78\Rightarrow x=39$

**15.** $(-3x)(2x-5)=-6x^2+15x$

**16.** $(-2x+1)+(5x-6)=-2x+1+5x-6$
$=3x-5$

**17.** $3.2x=19.2\Rightarrow x=\frac{19.2}{3.2}=6$

**18.** $0.12x+0.5x=1.86\Rightarrow x=\frac{1.86}{0.62}=3$

**19.** $0.75+0.15(7)=\$1.80$

**20.** $180.273=180.27,$ nearest hundredth

**21.** $20.17+13.59=33.76$

**22.** $27.01-5.3=21.71$

**23.** $(9.3)(10^2)=930$

**24.** $14.7\div 4.2=3.5$

**25.** $1.15(15)=\$17.25$

**26.** (a) $\frac{15}{25}=0.6$

(b) $\frac{15}{25}=0.6=60\%$

**27.** (a) $0.30=\frac{30}{100}=\frac{3}{10}$

(b) $0.30=\frac{30}{100}=30\%$

**28.** (a) $1\%=\frac{1}{100}$

(b) $1\%=\frac{1}{100}=0.01$

# Chapter 9

## 9.1 Exercises

**1.** Systems analysts

**3.** $3(50,000) = 150,000$

**5.** $9(50,000) - 3(50,000) = 300,000$

**7.** $13(17) = 221$ people/mi$^2$

**9.** $10(17) - 4(17) = 102$ more people

**11.** 31.76%

**13.** (a) Housing and transportation
(b) $31.76\% + 20.24\% = 52\%$

**15.** $0.3176(18,450) = \$5,859.72$

**17.** 17.1%

**19.** $10.4\% + 16.1\% = 26.5\%$

**21.** $100\% - 17.1\% = 82.9\%$

**23.** 42.7 in.

**25.** Anchorage, AK and Buffalo, NY

**27.** $63.3 - 31.9 = 31.4$ in.

**29.** Boston: $42.7 - 24.90 = 17.8$
Denver: $60.7 - 36.9 = 23.8$
Flagstaff: $109.1 - 66.5 = 42.6$
Great Falls: $63.3 - 31.9 = 31.4$
Omaha: $28.7 - 13.0 = 15.7$
Anchorage: $68.5 - 77.4 = -8.9$
Buffalo: $93.0 - 105.1 = -12.1$
Boise: $21.6 - 14.2 = 7.4$
Flagstaff, AZ with 42.6 in. has the greatest difference.

**31.** Flagstaff, AZ; Anchorage, AK; Buffalo, NY

**33.** June

**35.** (a) 4.5 thousand or 4500 customers
(b) Increased from 3500 to 4500

**37.**

| Month | Inc(+)/Dec(−) |
|---|---|
| Mar-Apr | $2000 - 3000 = -1000$ |
| Apr-May | $3500 - 2000 = 1500$ |
| May-Jun | $4500 - 3500 = 1000$ |
| Jun-Jul | $7000 - 4500 = 2500$ |
| Jul-Aug | $4000 - 7000 = -3000$ |

The largest increase in customers was between June and July, an increase of 2500.

**39.** 4 in.

**41.** April, May, and June

**43.** $4 - 1.5 = 2.5$ in.

**45.**

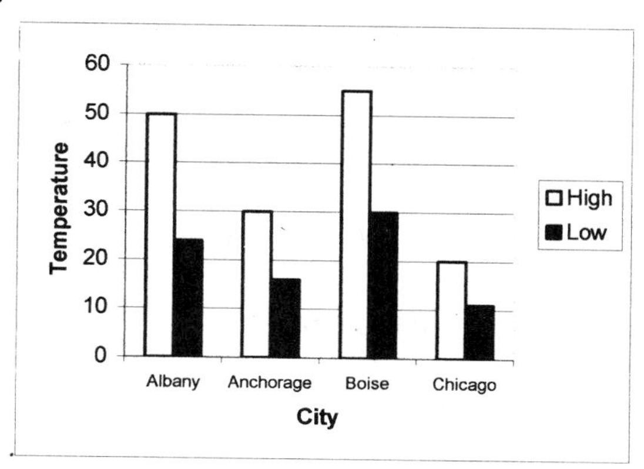

**47.**

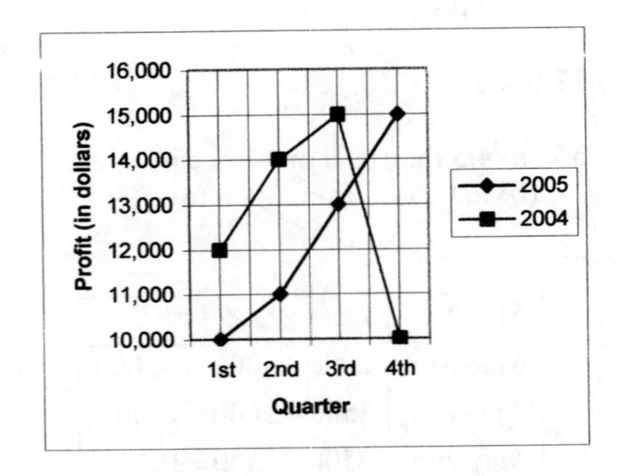


**49.**

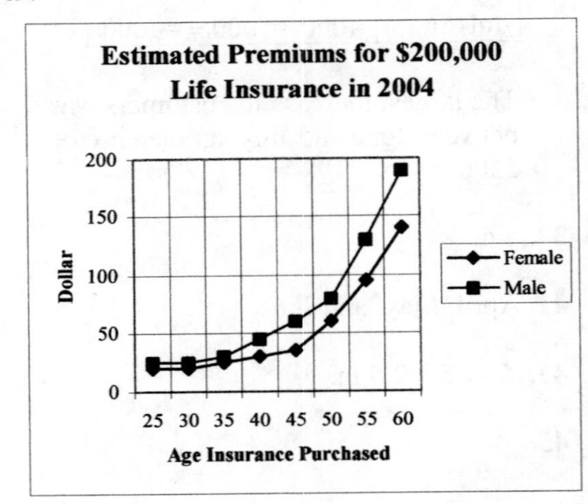


**51.** $100\% - 25.3\% = 74.7\%$

**53.** $0.175(100{,}000)(\$102) = \$1{,}785{,}000$

### Cumulative Review

**55.** $A = LW = 14(6) = 84 \text{ in.}^2$

**57.** $(2(7)(12) + 2(7)(20))\text{yd}^2 \cdot \dfrac{1 \text{ gal}}{28 \text{ yd}^2}$

$= 16 \text{ gal}$

### 9.2 Exercises

**1.** mean (average) of a set of values - the sum of the values divided by the number of values

**3.** $\dfrac{89+92+83+96+99}{5} = 91.8$

**5.** $\dfrac{2+3+3+4+2.5+2.5+4}{7} = 3 \text{ hr}$

**7.** $\dfrac{23+45+63+34+21+42}{6} = 38$

**9.** $\dfrac{189+193+162+102+189}{5} = 167$

$167,000

**11.** $\dfrac{6+5+4++5+8}{5} = 5.6 \text{ sales}$

**13.** $\dfrac{0+2+3+2+2}{5+4+6+5+4} = \dfrac{9}{24} = 0.375$

**15.** $\dfrac{276+350+391+336}{12+14+17+14}$

$= \dfrac{1353}{57} \approx 23.7 \text{ mi/gal}$

**17.** 22, 36, 45, 47, 48, 50, 58

median = 47

**19.** 865, 968, 999, 1023, 1052, 1152

median = $(999 + 1023) \div 2 = 1011$

**21.** 0.34, 0.52, 0.58, 0.69, 0.71

median = 0.58

**23.** 1.1, 1.9, 2.3, 2.9, 3.4, 3.9

$\text{median} = (2.3 + 2.9) \div 2 = 2.6$

**25.** \$11,600
\$15,700
\$17,000 ←
\$23,500
\$26,700
\$31,500

$\text{median} = (17,000 + 23,500) \div 2 = \$20,250$

**27.** 12, 20, 24, 26, 31, 40, 62, 108

$\text{median} = (26 + 31) \div 2 = 28.5 \text{ min}$

**29.** \$5.99
\$7.99
\$9.99 ←
\$11.99
\$13.99
\$17.99

$\text{median} = (9.99 + 11.99) \div 2 = \$10.99$

**31.** 22, 36, 36, 37, 44, 48, 53, 60, 64, 71

$\text{median} = (44 + 48) \div 2 = 46 \text{ actors}$

**33.** 59, $\underbrace{60, 60}_{\text{mode}}$, 65, 68, 72, 80

**35.** 116, 117, $\underbrace{121, 121}_{\text{mode}}$, 123, $\underbrace{150, 150}_{\text{mode}}$

**37.** $\underbrace{\$249, \$249}_{\text{mode}}$, \$259, \$269, \$439, \$649

**39.** 21, 42, 42, 45, 49, 55, 82

(a) $\dfrac{21 + 82 + 42 + 55 + 42 + 45 + 49}{7} = 48$

(b) median = 45 (c) mode = 42

**41.** 2.7, 6.1, 6.9, 7.1, 7.5

(a) $\dfrac{2.7 + 7.1 + 6.9 + 7.5 + 6.1}{5} = 6.06$

(b) median = 6.9 (c) mode - none

**43.** 73, 81, 81, 86, 92, 97

(a) $\dfrac{97 + 81 + 92 + + 73 + 86 + 81}{6} = 85$

(b) $\text{median} = (81 + 86) \div 2 = 83.5$

(c) mode = 81

**45.** (a)

$$\frac{1250 + 929 + 1399 + 990 + 1667 + 828 + 1276 + 1499}{8}$$

$= \dfrac{9838}{9} = 1229.75 \rightarrow \$1230$ to nearest dollar

(b) $1230 + 0.825 \times 1230 = \$1331.48$

**47.** $1331.48 + 128.03 = \$1459.51$

**49.** (a) $1500 + 1700 + 1650 + 1300 + 1440$
$+ 1580 + 1820 + 1380 + 2900 + 6300 = 21,570$

$21,570 \div 10 = \$2157$

(b) 1300
1380
1440
1500
1580 ←
1650
1700
1820
2900
6300

$\text{median} = (1580 + 1650) \div 2 = \$1615$

(c) Median, because mean is affected by the high salary of \$6300.

**51.** 1987, 2576, 3700, 4700, 5000, 7200, 8764, 9365

$$\text{median} = (4700 + 5000) \div 2 = 4850$$

## Cumulative Review

**53.** $\left.\frac{x}{2}+4\right|_{x=26} = \frac{26}{2}+4 = 13+4 = 17$

**55.** $\left.2x+1\right|_{x=5} = 2(5)+1 = 10+1 = 11$

**57.** $\frac{3}{8}\cdot 9200 = \$3450$

## How Am I Doing? Sections 9.1-9.2

**1.** 26%

**2.** Homes

**3.** $(0.35+0.19)70 = 37.8 \text{ mi}^2$

**4.** 4 in.

**5.** $5-3 = 2$ in.

**6.** January 2004

**7.** $\frac{4+3+3+5}{4} = 3.75$ in.

**8.** 7, 9, 14, 19, 25, 28, 32

↑

median = 19

**9.** 2, 5, 9, 13, 18, 23

↑

median = $(9+13) \div 2 = 11$

**10.** 79, 81, 83, $\underbrace{85,\ 85}_{\text{mode}}$

**11.** 4, 5, $\underbrace{7,\ 7}_{\text{mode}}$, $\underbrace{8,\ 8}_{\text{mode}}$, 9

**12.** 2.1, 8.7, 9.2, 9,6

↑ median = $(8.7+9.2) \div 2 = 8.95$

mode: none

$$\text{mean} = \frac{2.1+8.7+9.2+9.6}{4} = 7.4$$

## 9.3 Exercises

**1.** From origin, move 2 right and 1 down.

**3.** (1997,4700), (1998,4800), (1999,4950), (2000,4850)

**5.** Lena, (4,60); Janie, (7,62); Mark, (9,66)

**7.** (2,2)

**9.** (−1,4)

**11.** (3,−2)

**13.** (−2,−3)

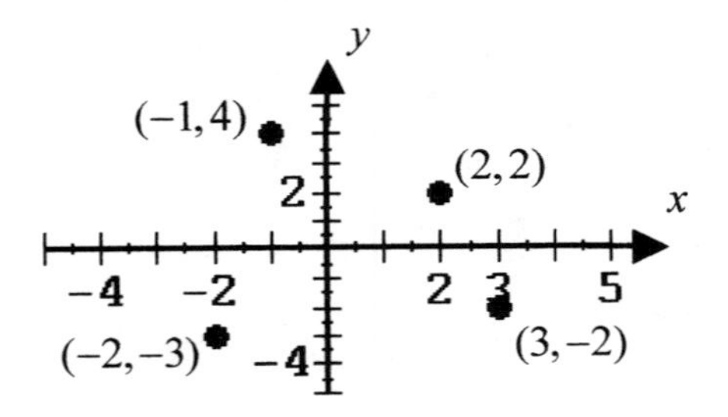

**15.** $(-1,2)$ **17.** $(5,-1)$ **19.** $(0,-2)$ **21.** $(5,0)$

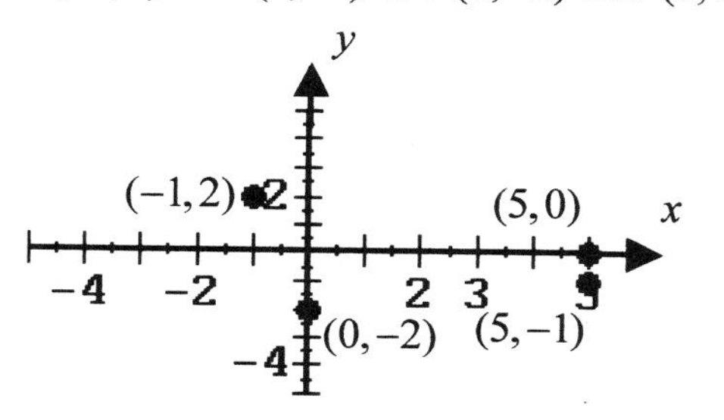


**23.** $\left(4\frac{1}{2},3\right)$ **25.** $\left(-1\frac{1}{2},-4\right)$ **27.** $\left(2\frac{1}{2},-1\right)$

**29.** $\left(-3\frac{1}{2},2\right)$

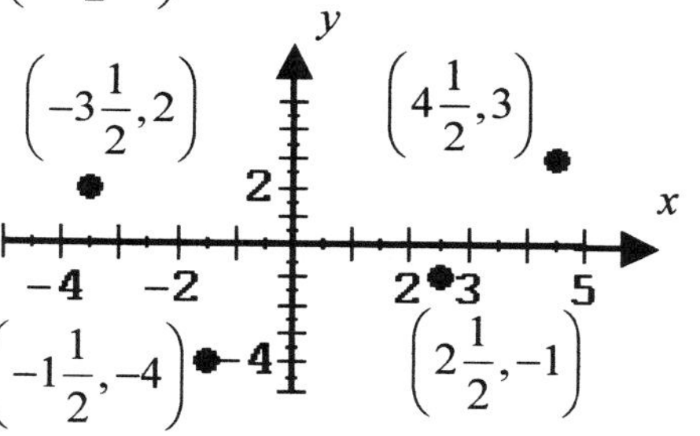


**31.** $K(-3,-3)$ **33.** $M(3,3)$

**35.** $O(-1,4)$ **37.** $Q\left(-3\frac{1}{2},4\right)$

**39.** $S\left(1\frac{1}{2},-3\right)$ **41.** $U(5,-1)$

**43.** (a) $(-4,-2)$

(b)

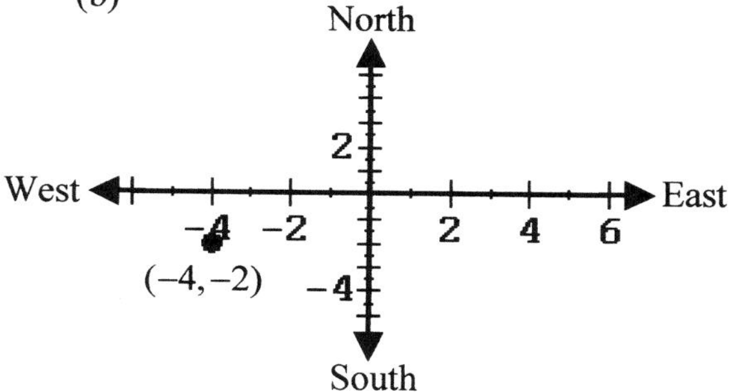


**45.** (a) $(2,1)$

(b)

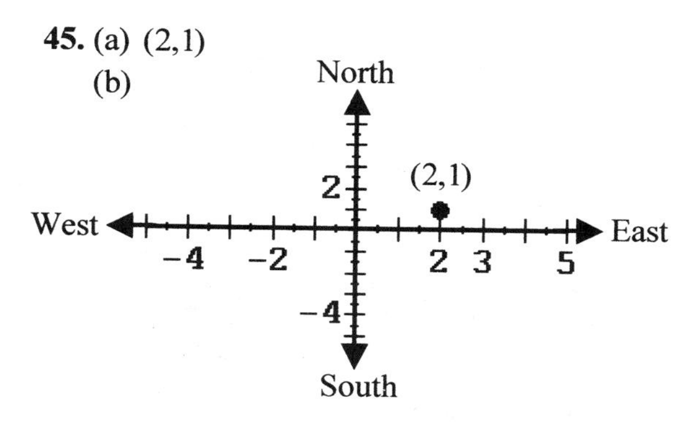


**47.** $(2,4)$, $(2,-1)$, $(2,-3)$, $(2,0)$

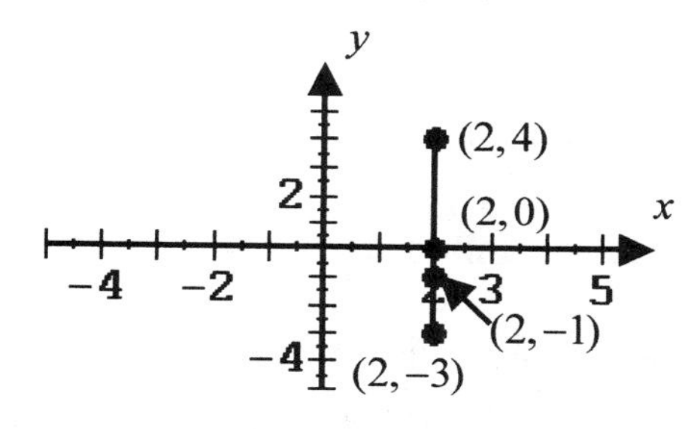


**49.** $(1,3)$, $(-5,3)$, $(0,3)$, $(-2,3)$

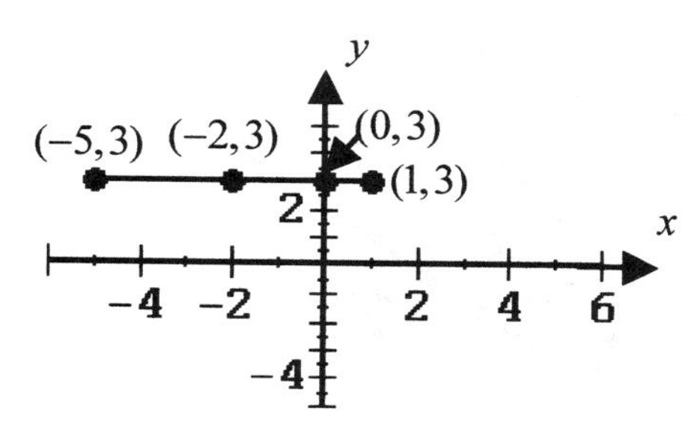


**51.** $(0,-1)$, $(-2,-1)$, $(-4,-1)$, $(4,-1)$

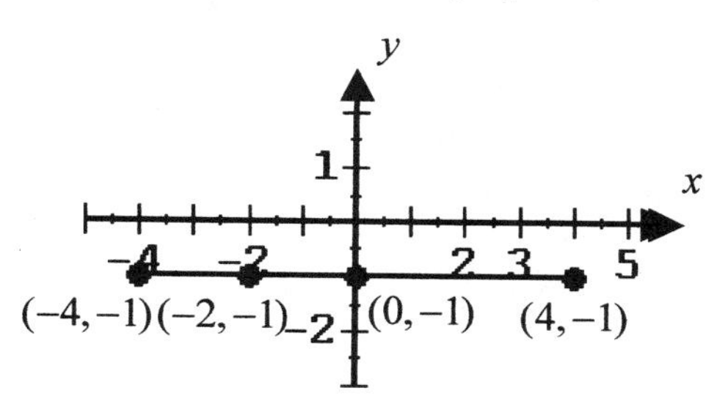

**53.** The figure is a rectangle.

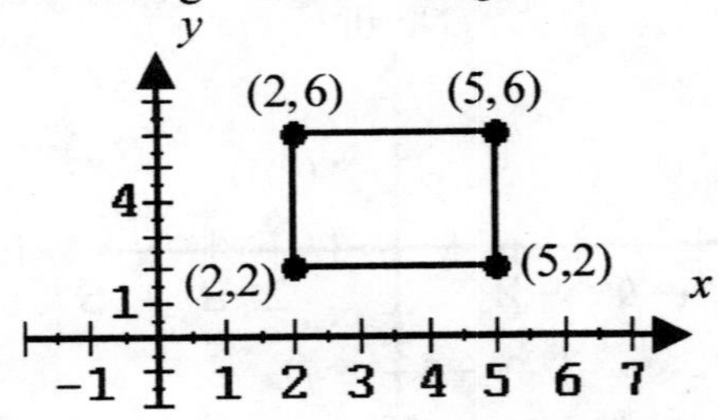


**Cumulative Review**

**55.** $4x-3|_{x=2}=4(2)-3=8-3=5$

**57.** $2x-6|_{x=1}=2(1)-6=-4$

**59.** 39(150 million km) = 5850 million km

**Putting Your Skills to Work, Problems for Individual Analysis**

**1.** (1,3), (2,6), (4,12), (6,18)

**2.**

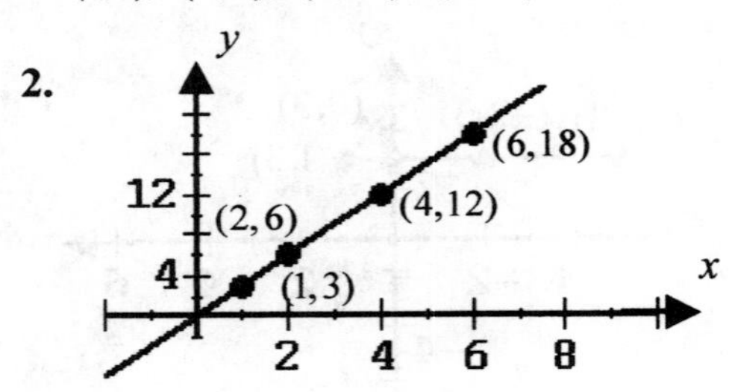


**3.** Positive correlation **4.** \$24

**5.** $P=3x$ where $x$ is the number of items

**6.** $P=3(20)=\$60$

**7.** (1,3), (2,4), (3,5), (4,6), (5,8)

**8.**

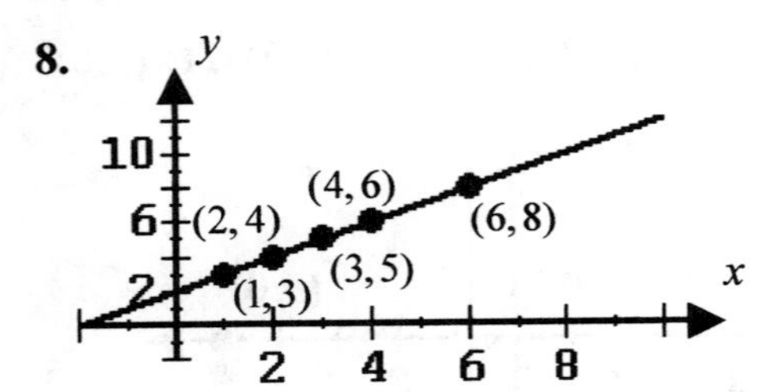


**9.** \$10

**10.** $C=x+2$ ($x=$ number of components)

**11.** $C=50+2=\$52$

**Putting Your Skills to Work, Problems for Group Investigation and Study**

**1.** (1,3), (2,6), (3,9), (4,12)

**3.** (1,2), (2,4), (3,6), (4,8)

**2. 4.**

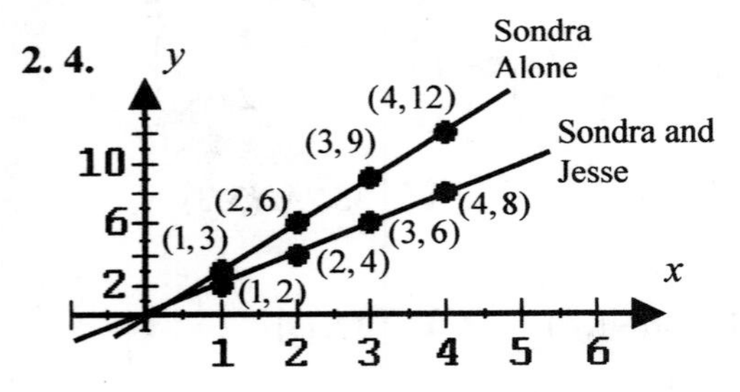


**5.** Sondra: $y=3x$; together: $y=2x$

**6.** (a) $y=12(3)=36$ hr, Sondra alone
(b) $y=12(2)=24$ hr, Together

**7.** $36(20)=\$720$ for Sondra alone
$24(20)+24(15)=\$840$ together
It would cost less for Sondra alone.

**8.** Negative values make no sense here.
$(-1,5)$, $(0,4)$, $(1,3)$, $(4,0)$ are solutions.

**9.4 Exercises**

**1.** For (2,5), $x+2y=2+2(5)=12\neq 4$ so (2,5) is not a solution.

**3.** $x+y=4\Rightarrow y=4-x$
$y=4-(-1)=5\Rightarrow(-1,5)$
$y=4-0=4\Rightarrow(0,4)$
$y=4-1=3\Rightarrow(1,3)$
$y=4-4=0\Rightarrow(4,0)$

**5.** $x+y=12\Rightarrow y=12-x$
$y=12-(-1)=13\Rightarrow(-1,13)$
$y=12-0=12\Rightarrow(0,12)$
$y=12-1=11\Rightarrow(1,11)$
$y=12-12=0\Rightarrow(12,0)$
$(-1,13)$, $(0,12)$, $(1,11)$, $(0,12)$ are solutions to $x+y=12$.

**7.** $y=35x$

(a) $140=35x\Rightarrow x=\dfrac{140}{35}=4,\ (4,140)$

(b) $280=35x\Rightarrow x=\dfrac{280}{35}=8,\ (8,280)$

**9.** $y=80x$
(a) $240=80x\Rightarrow x=3\Rightarrow(3,240)$
(b) $400=80x\Rightarrow x=5\Rightarrow(5,400)$

**11.** $x+2y=16\Rightarrow y=\dfrac{16-x}{2}$

$y=\left.\dfrac{16-x}{2}\right|_{x=0}=8\Rightarrow(0,8)$

$x+2(0)=16\Rightarrow x=16\Rightarrow(16,0)$
$x+2(4)=16\Rightarrow x=8\Rightarrow(8,4)$

**13.** $x+y=5$
$x+2=5\Rightarrow x=3\Rightarrow(3,2)$
$0+y=5\Rightarrow y=5\Rightarrow(0,5)$
$1+y=5\Rightarrow y=4\Rightarrow(1,4)$

**15.** $y=x+2$
$y=-1+2=1\Rightarrow(-1,1)$
$3=x+2\Rightarrow x=1\Rightarrow(1,3)$
$0=x+2\Rightarrow x=-2\Rightarrow(-2,0)$

**17.** $y=5x+3$
$y=5(0)+3=3\Rightarrow(0,3)$
$y=5(-1)+3=-2\Rightarrow(-1,-2)$
$y=5(1)+3=8\Rightarrow(1,8)$

**19.** $y=5x-3$
$y=5(0)-3=-3\Rightarrow(0,-3)$

$y=5\left(\dfrac{3}{5}\right)-3=0\Rightarrow\left(\dfrac{3}{5},0\right)$

$y=5(1)-3=2\Rightarrow(1,2)$

**21.** $y=x+6$
$y=0+6=6\Rightarrow(0,6)$
$y=-6+6=0\Rightarrow(-6,0)$
$y=1+6=7\Rightarrow(1,7)$

**23.** $y=2x+2$
$y=2(-2)+2=-2\Rightarrow(-2,-2)$
$y=2(-1)+2=0\Rightarrow(-1,0)$
$y=2(0)+2=2\Rightarrow(0,2)$

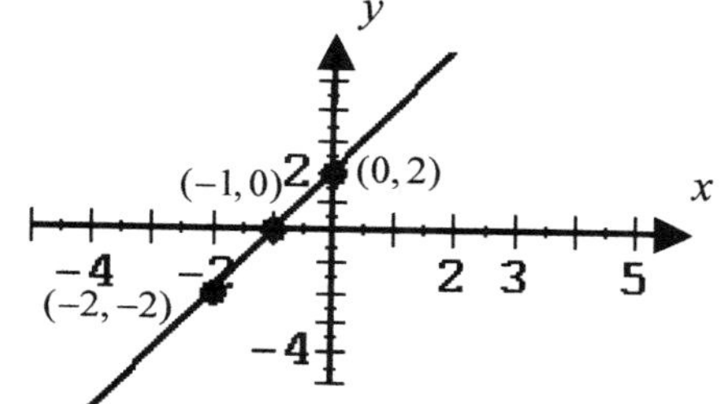

**25.** $y = -3x + 1,\ y = -3(-1) + 1 = 4 \Rightarrow (-1, 4)$
$y = -3(0) + 1 = 1 \Rightarrow (0, 1)$
$y = -3(1) + 1 = -2 \Rightarrow (1, -2)$

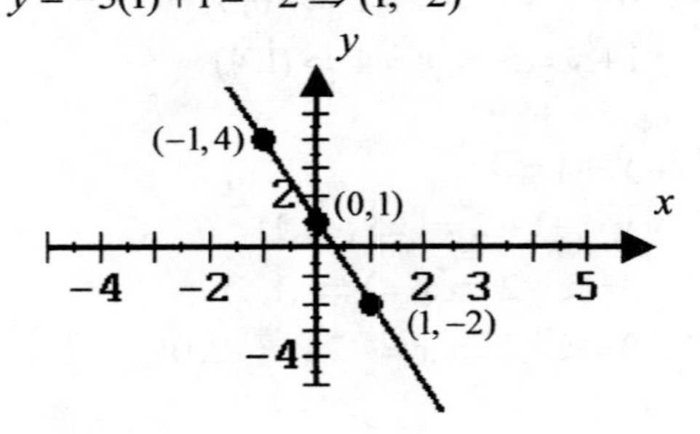


**27.** $y = 5x - 4$
$y = 5(0) - 4 = -4 \Rightarrow (0, -4)$
$y = 5(1) - 4 = 1 \Rightarrow (1, 1)$
$y = 5\left(\frac{2}{5}\right) - 4 = -2 \Rightarrow \left(\frac{2}{5}, -2\right)$

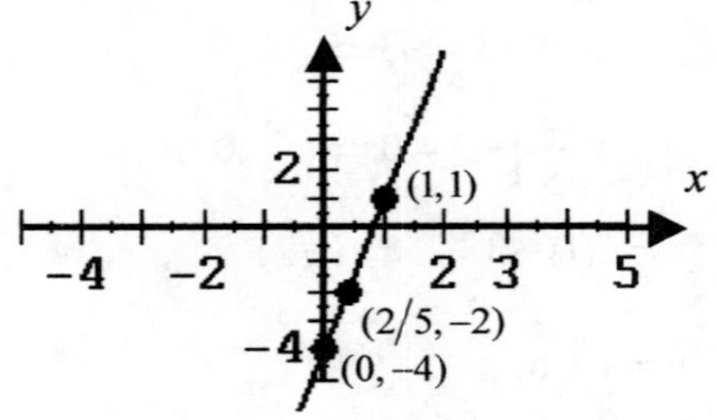


**29.** $y = 3x - 2$
$y = 3(-1) - 2 = -5 \Rightarrow (-1, -5)$
$y = 3(0) - 2 = -2 \Rightarrow (0, -2)$
$y = 3(1) - 2 = 1 \rightarrow\Rightarrow (1, 1)$

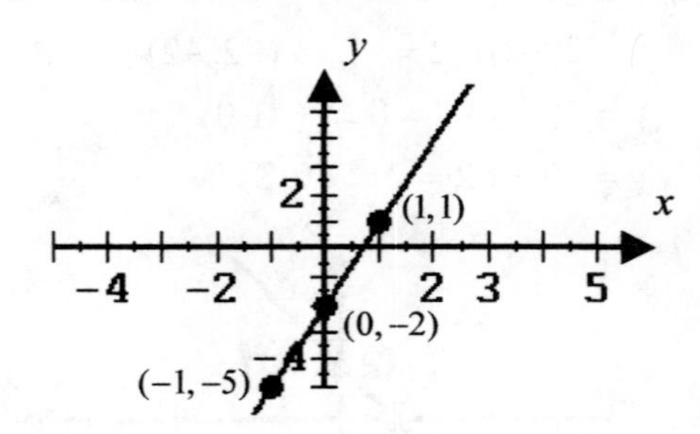


**31.** $y = -5x - 7$
$y = -5(-2) - 7 = 3 \Rightarrow (-2, 3)$
$y = -5\left(-\frac{3}{2}\right) - 7 = \frac{1}{2} \Rightarrow \left(-\frac{3}{2}, \frac{1}{2}\right)$
$y = -5(-1) - 7 = -2 \Rightarrow (-1, -2)$

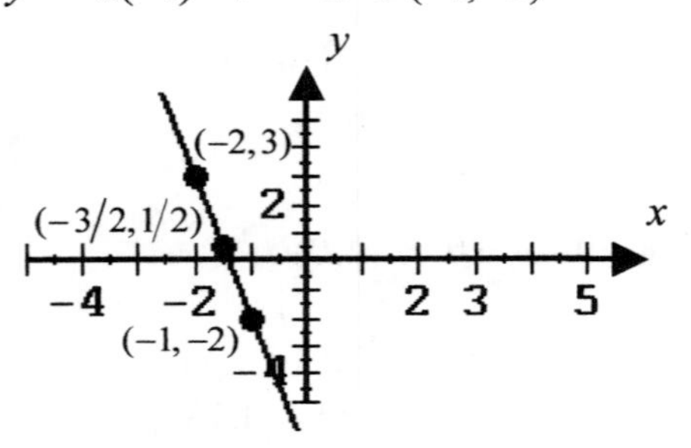


**33.** $y = 3 \Rightarrow (-4, 3),\ (0, 3),\ (5, 3)$

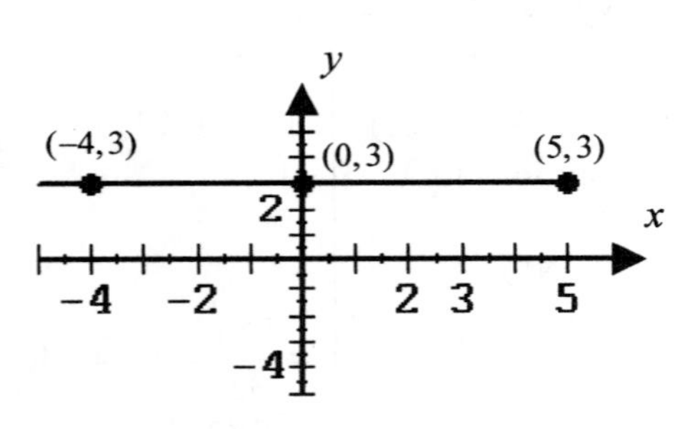


**35.** $x = -2 \Rightarrow (-2, 4),\ (-2, 0),\ (-2, -3)$

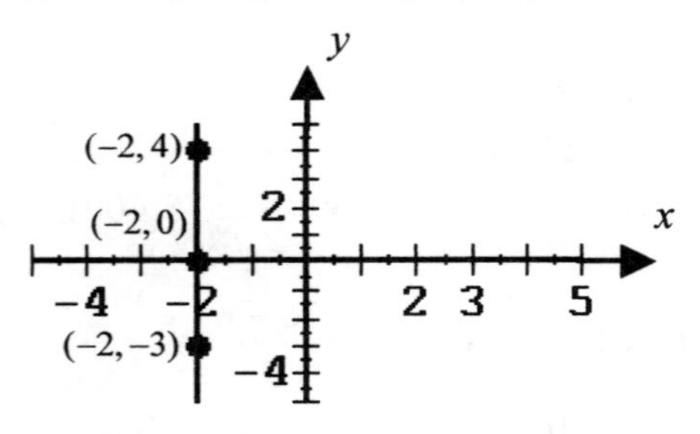


**37.** (6, 0) is not a solution because it is not on the line formed by the other three points.

**39.** $(2, 4) = (2, 2 \cdot 2),\ (3, 6) = (3, 2 \cdot 3)$
$(4, 8) = (4, 2 \cdot 4) \Rightarrow y = 2x$

## Cumulative Review

**41.** $2x+3=10 \Rightarrow 2x=7 \Rightarrow x=\frac{7}{2}$

**43.** $-3x-5=14 \Rightarrow -3x=19 \Rightarrow x=-\frac{19}{3}$

**45.** $\frac{\$8}{\text{hr}}\cdot 25\frac{1}{2}\text{ hr}=\$204$

## Chapter 9 Review Problems

**1.** $7(50)=350$

**2.** $6(50)=300$

**3.** $9(50)-300=150$

**4.** $350+450=800$

**5.** 15%

**6.** 4%

**7.** $27\%+30\%=57\%$

**8.** $15\%+8\%+4\%=27\%$

**9.** $0.08(2400)=\$192$

**10.** $0.15(2400)=\$360$

**11.** $(0.30+0.04)(2400)=\$816$

**12.** $(0.15+0.27)(2400)=\$1008$

**13.** (a) China with 1.2 billion
(b) India with 1.6 billion

**14.** (a) Brazil with 0.170 billion
(b) Brazil with 0.244 billion

**15.** Third largest behind China and India.

**16.** Fourth largest behind China, India, and the United States.

**17.** $1.3-1.2=0.1$ billion more people in 2050

**18.** $0.403-0.275=0.128$ billion more people in 2050

**19.** China: 0.1; India: 0.6; Indonesia: 0.1; United States: 0.128; Brazil: 0.074
India has the largest increase with 0.6 billion.

**20.** 0.1 billion more people in 2050

**21.** (a) \$400 (b) \$350

**22.** (a) \$350 (b) \$300

**23.**

| Item | Loss |
|---|---|
| Antilock Brakes | 50 |
| Auto. Trans. | 50 |
| CD Changer | 25 |
| Alarm System | 25 |
| Air Conditioning | 50 |
| Alum. Wheels | 25 |

Antilock brakes, automatic transmission, and air conditioning lost the most value, \$50.

**24.** CD changer, alarm system, and aluminum wheels lost the least value, \$25.

**25.** $\frac{300+350+250+175+400+175}{6}$

$=\frac{1650}{6}$

$=\$275$

**26.** $\dfrac{250+300+225+150+350+150}{6}$

$= \dfrac{1425}{6}$

$= \$237.50$

**27.** 175, 175, 250, 300, 350, 400

↑

median $= (250+300) \div 2 = \$275$

**28.** 150, 150, 225, 250, 300, 350

↑

median $= (225+250) \div 2 = \$237.50$

**29.**

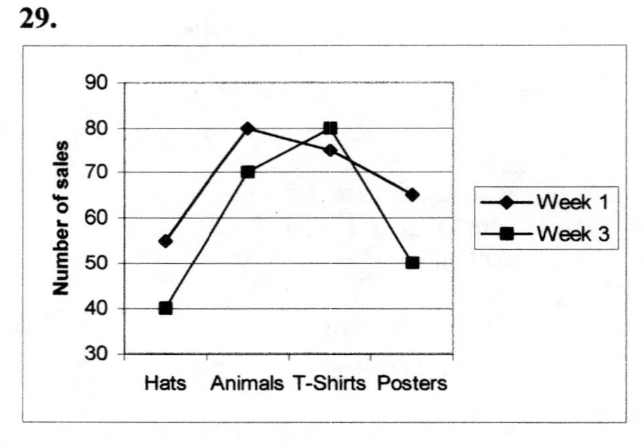


**30.**

**31.**

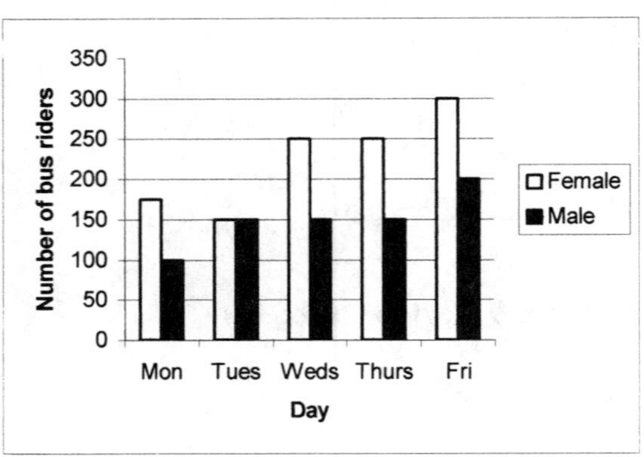


**32.**

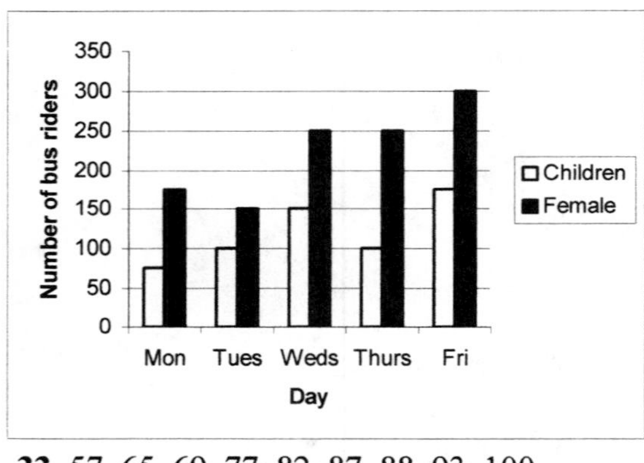


**33.** 57, 65, 69, 77, 82, 87, 88, 93, 100

↑

median = 82

**34.** 58, 77, 79, 81, 83, 87, 88, 91, 104

↑

median = 83

**35.** 0,1,4,5,9,18,19,19,20,21,22,25,27,36,38,43

↑

median $= \dfrac{19+20}{2} = 19.5$

**36.** 0,3,9,13,14,15,16,18,19,21,24,25,26,28,31,36

$$\text{median} = \frac{18+19}{2} = 18.5$$

**37.** $\frac{86+83+88+95+97+100+81}{7} = 90°\text{F}$

**38.** $\frac{87+105+89+120+139+160+98}{7}$

$= \$114$

**39.** $\frac{76+20+91+57+42+21+75+82}{8} = 58$

**40.** $151+140+148+156+183+201$
$+205+228+231+237 = 1880$

$\frac{1880}{10} = 188$

**41.** 13, 13, 14, 18, 19, 22
(13, 13: mode)

**42.** 14, 14, 18, 18, 18, 28, 29
(18, 18, 18: mode)

**43.** (1997,5000), (1998,6000), (1999,5500), (2000,6250)

**44.** (2001,7000), (2002,6500), (2003,7500), (2004,7250)

**45. 46. 47. 48.** See graph below.

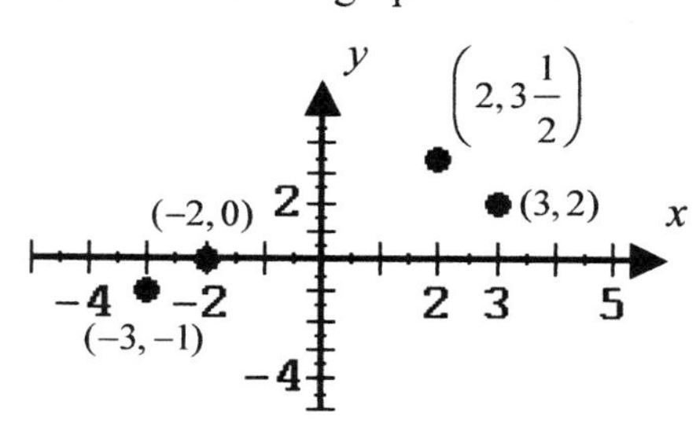


**49.** $R(-5,3)$

**50.** $S(-1,-1)$

**51.** $T\left(0, 2\frac{1}{2}\right)$

**52.** $U(4,1)$

**53.** (2,0), (2,3), (2,−1)

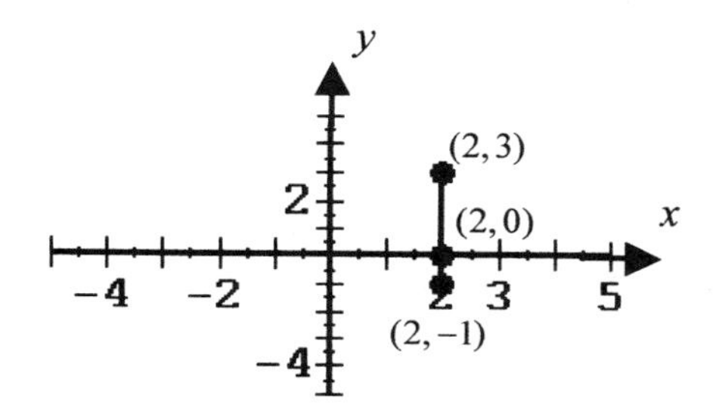


**54.** (3,1), (−4,1), (0,1)

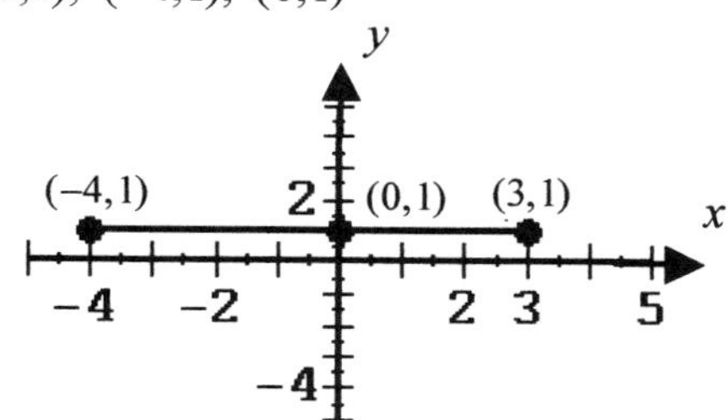


**55.** $x + y = 8 \Rightarrow y = 8 - x$
$y = 8 - 6 = 2 \Rightarrow (6,2)$
$y = 8 - 8 = 0 \Rightarrow (8,0)$
$y = 8 - 7 = 1 \Rightarrow (7,1)$

**56.** $x + y = 3 \Rightarrow y = 3 - x$
$y = 3 - 0 = 3 \Rightarrow (0,3)$
$y = 3 - 2 = 1 \Rightarrow (2,1)$
$y = 3 - 3 = 0 \Rightarrow (3,0)$

**57.** (a) $y = 70x \Rightarrow 280 = 70x \Rightarrow x = 4$
$(4, 280)$
(b) $y = 70x \Rightarrow 350 = 70x \Rightarrow x = 5$
$(5, 350)$

**58.** $y = 2x - 6 \Rightarrow x = \dfrac{y+6}{2}$

$x = \dfrac{-4+6}{2} = 1 \Rightarrow (1, -4)$

$x = \dfrac{-6+6}{2} = 0 \Rightarrow (0, -6)$

$x = \dfrac{-8+6}{2} = -1 \Rightarrow (-1, -8)$

**59.** $y = -6x + 2 \Rightarrow x = \dfrac{2-y}{6}$

$x = \dfrac{2-2}{6} = 0 \Rightarrow (0, 2)$

$x = \dfrac{2-8}{6} = -1 \Rightarrow (-1, 8)$

$x = \dfrac{2-(-4)}{6} = 1 \Rightarrow (1, -4)$

**60.** $y = 3x - 1\big|_{x=-1} = -4 \Rightarrow (-1, -4)$

$y = 3x - 1\big|_{x=0} = -1 \Rightarrow (0, -1)$

$y = 3x - 1\big|_{x=1} = 2 \Rightarrow (1, 2)$

**61.** $y = -5x - 4\big|_{x=-\frac{3}{2}} = -\dfrac{7}{2} \Rightarrow \left(-\dfrac{3}{2}, -\dfrac{7}{2}\right)$

$y = -5x - 4\big|_{x=-1} = 1 \Rightarrow (-1, 1)$

$y = -5x - 4\big|_{x=0} = -4 \Rightarrow (0, -4)$

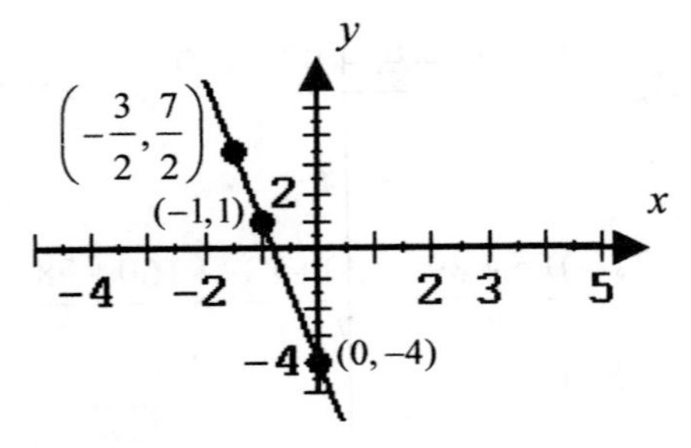


**62.** $y = 4x - 6\big|_{x=1} = -2 \Rightarrow (1, -2)$

$y = 4x - 6\big|_{x=\frac{3}{2}} = 0 \Rightarrow \left(\dfrac{3}{2}, 0\right)$

$y = 4x - 6\big|_{x=2} = 2 \Rightarrow (2, 2)$

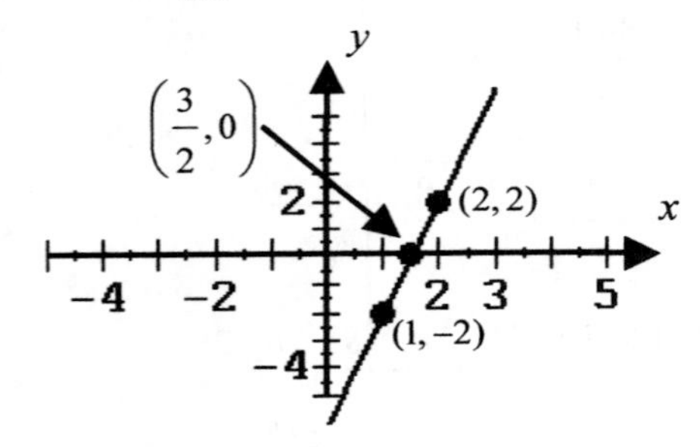


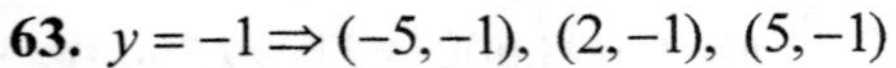
**63.** $y = -1 \Rightarrow (-5, -1),\ (2, -1),\ (5, -1)$

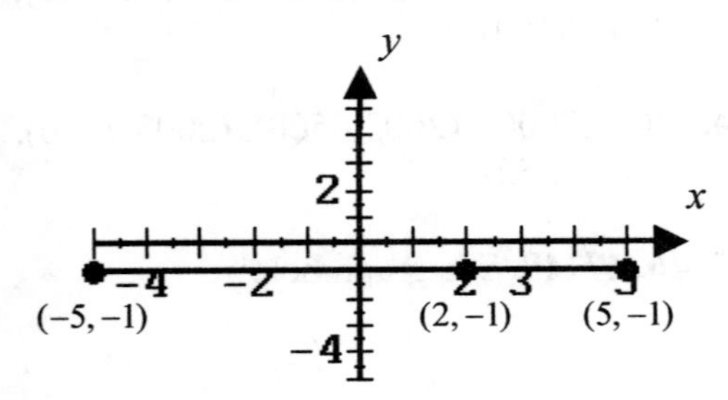

**How Am I Doing? Chapter 9 Test**

**1.** Age 18-20, 44%.

**2.** $44\% + 33\% = 77\%$

**3.** $44\% + 6\% = 50\%$

**4.** $0.07(5000) = 350$ students

**5.** 11; 12

**6.** 6; 9

**7.** Otsego Lake, NY

**8.** Baikal Lake, Russia

**9.** Honda and Toyota do not have a road-side assistance plan.

**10.** 5 yr

**11.** 3 more yr

**12.** (a) V.W. Passat and Lexus ES 330 (b) Lexus ES 330 (c) V.W. Passat (d) V.W. Passat and Lexus

**13.** Lexus: $\dfrac{4+6+6+4}{4} = 5$ yr

Honda: $\dfrac{3+3+5+0}{4} = 2.75$ yr

**14.** Toyota: $\dfrac{3+5+5+0}{4} = 3.25$ yr

Volkswagen: $\dfrac{4+5+12+4}{4} = 6.25$ yr

**15.** $\dfrac{89+46+85+91+83+90}{6} = 80.\overline{6}$

**16.** 76, 83, 85, 89, 90, 91

$$\text{median} = \frac{85+89}{2} = 87$$

**17.** $(3, 5)$ See graph below.

**18.** $(0, 0)$ See graph below.

**19.** $(2, -1)$ See graph below.

**20.** $(-3, 0)$ See graph below.

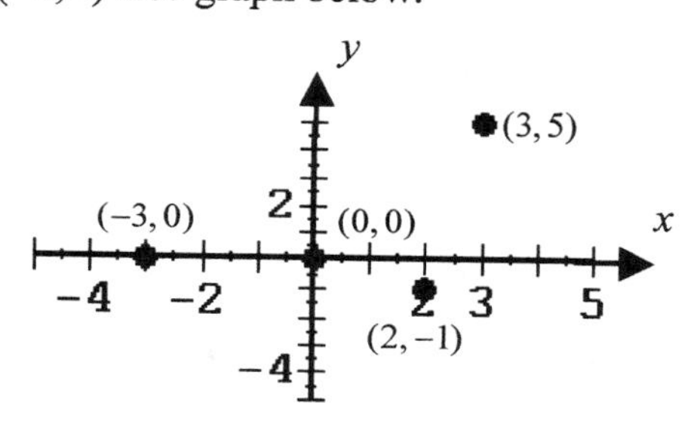


**21.** $A(4, 2)$

**22.** $B(0, -2)$

**23.** $C(3, -1)$

**24.** $D(-1, -3)$

**25.** $(1, 2)$, $(-3, 2)$, $(0, 2)$, $(4, 2)$

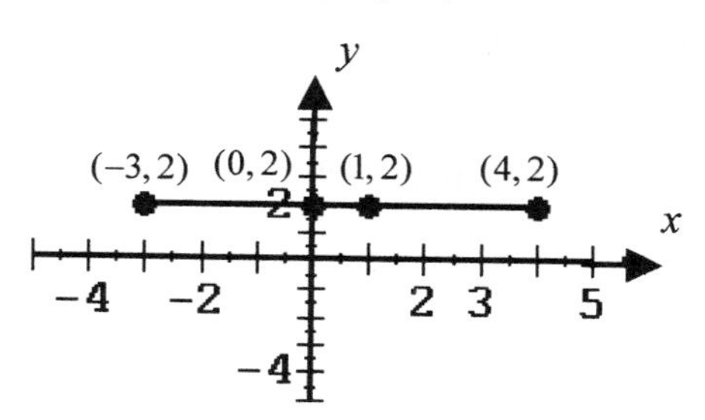

**26.** $y = 4x - 2\big|_{x=0} = -2 \Rightarrow (0, -2)$

$y = 4x - 2\big|_{x=\frac{1}{2}} = 0 \Rightarrow \left(\frac{1}{2}, 0\right)$

$y = 4x - 2\big|_{x=1} = 2 \Rightarrow (1, 2)$

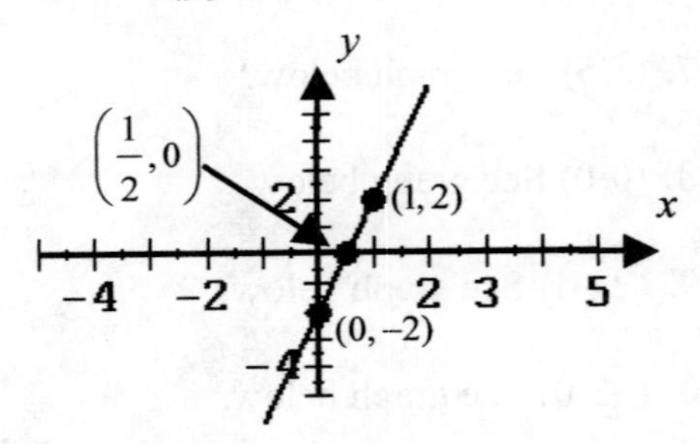


**27.** $y = 3 \Rightarrow (-3, 3),\ (0, 3),\ (3, 3)$

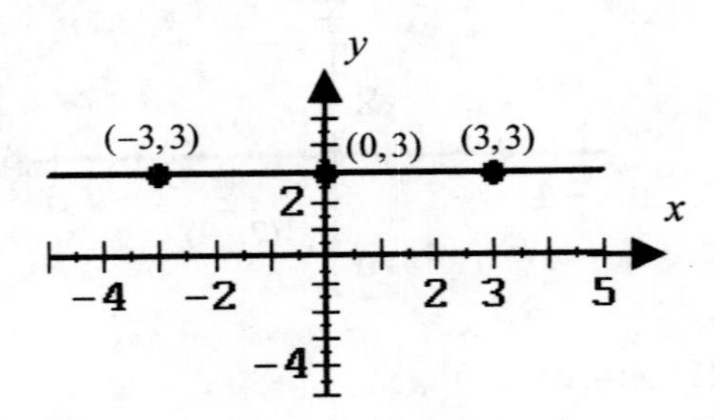


**Cumulative Test for Chapters 1-9**

**1.** $5 \cdot 5 \cdot 5 \cdot 5 \cdot 5 \cdot 5 = 5^6$

**2.** $4^3 = 4 \cdot 4 \cdot 4 = 64$

**3.** $3^2 - 2 + 7 = 9 + 5 = 14$

**4.** $20 \div 4 + 5(2+1) = 5 + 5(3) = 5 + 15 = 20$

**5.** $-2^3 = -2 \cdot 2 \cdot 2 = -8$

**6.** $98 = 2 \cdot 7 \cdot 7 = 2 \cdot 7^2$

**7.** $\dfrac{285 \text{ mi}}{12 \text{ gal}} \approx 23.8 \text{ mi/gal}$

**8.** $\dfrac{xy^2}{18} \cdot \dfrac{9}{x} = \dfrac{y^2}{2}$

**9.** The reciprocal of $-11$ is $\dfrac{1}{-11} = -\dfrac{1}{11}$

**10.** $\dfrac{2}{7} \div \dfrac{-8}{14} = -\dfrac{2}{7} \cdot \dfrac{14}{8} = -\dfrac{1}{2}$

**11.** $10x = 120 \Rightarrow x = 12$

check: $10(12) \stackrel{?}{=} 120,\ 120 = 120$

**12.** $\dfrac{x}{3} = 12 \Rightarrow x = 3(12) = 36$

check: $\dfrac{36}{3} \stackrel{?}{=} 12,\ 12 = 12$

**13.** $-5x - 2 = -22 \Rightarrow 5x = 20 \Rightarrow x = 4$

check: $-5(4) - 2 \stackrel{?}{=} -22,\ -22 = -22$

**14.** $2(2x^3 - x + 6) = 4x^3 - 2x + 12$

**15.** $(3x^2 - 2x - 8) + (x^2 + 4x - 3)$
$= 3x^2 - 2x - 8 + x^2 + 4x - 3$
$= 4x^2 + 2x - 11$

**16.** $(3x + 1) - (5x - 2) = 3x + 1 - 5x + 2$
$= -2x + 3$

**17.** 20%

**18.** 10%

**19.** $60\% - 5\% = 55\%$

**20.** $45\% - 15\% = 30\%$

**21.** $\dfrac{76+83+72+86+88}{5}=81$

**22.** 68, 80, 90, 105, 115, 121

$\text{median}=\dfrac{90+105}{2}=97.5$

**23.** 1, 2, 6, 7, $\underbrace{8,\ 8}_{\text{mode}}$, 9

**24.** $(6,1)$ See graph below.

**25.** $(2,3)$ See graph below.

**26.** $(0,-2)$ See graph below.

**27.** $(-4,-3)$ See graph below.

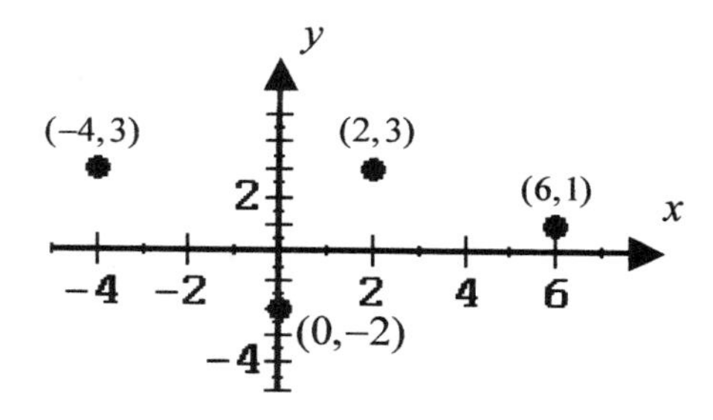


**28.** $y=x+3\big|_{x=-1}=2\Rightarrow(-1,2)$

$y=x+3\big|_{x=0}=3\Rightarrow(0,3)$

$y=x+3\big|_{x=1}=4\Rightarrow(1,4)$

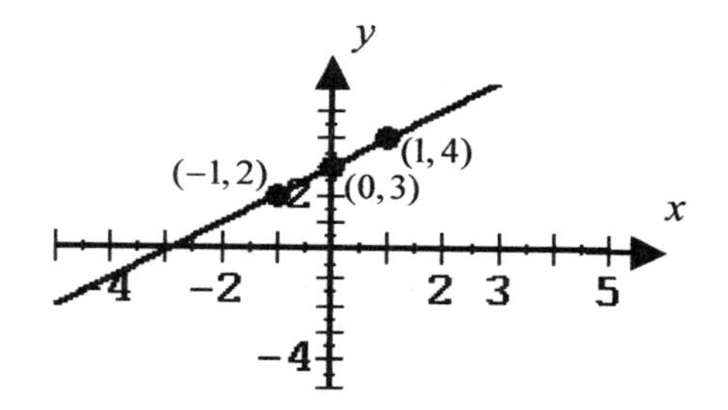


**29.** $y=-3x-4\big|_{x=-2}=2\Rightarrow(-2,2)$

$y=-3x-4\big|_{x=-1}=-1\Rightarrow(-1,-1)$

$y=-3x-4\big|_{x=0}=-4\Rightarrow(0,-4)$

# Chapter 10

## 10.1 Exercises

**1.** 1 ft = 12 in.

**3.** 2 pints = 1 quart

**5.** 1 ton = 2000 lb

**7.** 4 qt = 1 gal

**9.** 7 days = 1 week

**11.** 60 sec = 1 min

**13.** $15 \text{ ft} \cdot \frac{\text{yd}}{3 \text{ ft}} = 5 \text{ yd}$

**15.** $10,560 \text{ ft} \cdot \frac{1 \text{ mi}}{5280 \text{ ft}} = 2 \text{ mi}$

**17.** $87 \text{ in.} \cdot \frac{1 \text{ ft}}{12 \text{ in}} = 7.25 \text{ ft}$

**19.** $13 \text{ tons} \cdot \frac{2000 \text{ lb}}{1 \text{ ton}} = 26,000 \text{ lb}$

**21.** $7 \text{ gal} \cdot \frac{4 \text{ qt}}{1 \text{ gal}} = 28 \text{ qt}$

**23.** 11 days = 264 hours

**25.** $218 \text{ ft} \cdot \frac{12 \text{ in.}}{\text{ft}} + 10 \text{ in.} = 2626 \text{ in.}$

**27.** $\frac{\$2.25}{\text{hr}} \cdot \frac{24 \text{ hr}}{\text{day}} \cdot 2\frac{1}{2} \text{day} = \$135$

**29.** $\frac{\$3.00}{\text{lb}} \cdot \frac{\text{lb}}{16 \text{ oz}} \cdot 24 \text{ oz} = \$4.50$

**31.** 37 cm = 370 mm

**33.** 3.6 km = 3600 m

**35.** 2.43 kL = 2,430,000 mL

**37.** 28,156 mL = 0.028156 kL

**39.** 0.78 g = 0.00078 kg

**41.** 5.9 kg = 5,900,000 mg

**43.** 7 mL = 0.007 L = 0.000007 kL

**45.** 522 mg = 0.522 g = 0.000522 kg

**47.** 35 mm = 3.5 cm = 0.035 m

**49.** 3582 mm = 3.582 m = 0.003582 km

**51.** 0.32 cm = 0.0032 m = 0.0000032 km

**53.** $\frac{\$6.00}{\text{mL}} \cdot \frac{1000 \text{ mL}}{\text{L}} \cdot 1 \text{ L} = \$6000$

**55.** $\frac{\$850}{\text{mL}} \cdot \frac{1000 \text{ mL}}{\text{L}} \cdot 0.4 \text{ L} = \$340,000$

**57.** (a) 4818 m = 481,800 cm
(b) 4818 m = 4.818 km

**59.** $5280 \text{ ft} \cdot \frac{1 \text{ yd}}{3 \text{ ft}} = 1760 \text{ yd}$

**61.** $6 \text{ mi} \cdot \frac{5280 \text{ ft}}{\text{mi}} = 31,680 \text{ ft}$

**63.** $6 \text{ tons} \cdot \frac{2000 \text{ lb}}{1 \text{ ton}} = 12,000 \text{ lb}$

**65.** 14.6 kg = 14,600 g

**67.** $3.22 \text{ kL} = 3220 \text{ L}$

**69.** $7183 \text{ mg} = 7.183 \text{ g} = 0.007183 \text{ kg}$

**71.** $3 \text{ yr} \cdot \frac{365 \text{ day}}{1 \text{ yr}} \cdot \frac{24 \text{ hr}}{1 \text{ day}} = 26,280 \text{ hr}$

**73.** $528 \text{ megabytes} = 528,000,000 \text{ bytes}$

**75.** $24.9 \text{ gigabytes} = 24,900,000,000 \text{ bytes}$

**Cumulative Review**

**77.** $a = 0.23(250) = 57.5$

**79.** $0.08(8960) = \$716.80$

**10.2 Exercises**

**1.** $7 \text{ ft} \cdot \frac{0.305 \text{ m}}{\text{ft}} \approx 2.14 \text{ m}$

**3.** $14 \text{ m} \cdot \frac{1.09 \text{yd}}{1 \text{ m}} = 15.26 \text{ yd}$

**5.** $15 \text{ km} \cdot \frac{0.62 \text{ mi}}{1 \text{ km}} = 9.3 \text{ mi}$

**7.** $24 \text{ yd} \cdot \frac{0.914 \text{ m}}{\text{yd}} \approx 21.94 \text{ m}$

**9.** $82 \text{ mi} \cdot \frac{1.61 \text{ km}}{1 \text{ mi}} = 132.02 \text{ km}$

**11.** $25 \text{ m} \cdot \frac{3.28 \text{ ft}}{\text{m}} = 82 \text{ ft}$

**13.** $17.5 \text{ cm} \cdot \frac{0.394 \text{ in.}}{\text{cm}} \approx 6.90 \text{ in.}$

**15.** $5 \text{ gal} \cdot \frac{3.79 \text{ L}}{1 \text{ gal}} = 18.95 \text{ L}$

**17.** $4.5 \text{ L} \cdot \frac{1.06 \text{ qt}}{1 \text{ L}} = 4.77 \text{ qt}$

**19.** $7 \text{ oz} \cdot \frac{28.35 \text{ g}}{\text{oz}} = 198.45 \text{ g}$

**21.** $16 \text{ kg} \cdot \frac{2.2 \text{ lb}}{\text{kg}} = 35.2 \text{ lb}$

**23.** $126 \text{ g} \cdot \frac{0.0353 \text{ oz}}{\text{g}} \approx 4.45 \text{ oz}$

**25.** $4 \text{ kg} \cdot \frac{2.2 \text{ lb}}{\text{kg}} \cdot \frac{16 \text{ oz}}{1 \text{ lb}} = 140.8 \text{ oz}$

**27.** $166 \text{ cm} \cdot \frac{1 \text{ in.}}{2.54 \text{ cm}} \cdot \frac{1 \text{ ft}}{12 \text{ in.}} = 5.45 \text{ ft}$

**29.** $16.5 \text{ ft} \cdot \frac{12 \text{ in}}{\text{ft}} \cdot \frac{2.54 \text{ cm}}{\text{in.}} = 502.92 \text{ cm}$

**31.** $50 \frac{\text{km}}{\text{hr}} \cdot \frac{0.62 \text{ mi}}{\text{km}} = 31 \frac{\text{mi}}{\text{hr}}$

**33.** $60 \frac{\text{mi}}{\text{hr}} \cdot \frac{1.61 \text{ km}}{\text{mi}} = 96.6 \frac{\text{km}}{\text{hr}}$

**35.** $13 \text{ mm} \cdot \frac{\text{cm}}{10 \text{ mm}} \cdot \frac{1 \text{ in.}}{2.54 \text{ cm}} \approx 0.51 \text{ in.}$

**37.**
$$\begin{aligned} F = 1.8C + 32 &= 1.8(40) + 32 \\ &= 104 \text{ °F} \end{aligned}$$

**39.**
$$\begin{aligned} F = 1.8C + 32 &= 1.8(85) + 32 \\ &= 185 \text{ °F} \end{aligned}$$

**41.** $C = \dfrac{5F - 160}{9}$

$C = \dfrac{5(168) - 160}{9} \approx 75.56°C$

**43.** $C = \dfrac{5F - 160}{9}$

$C = \dfrac{5(86) - 160}{9} = 30°C$

**45.** $9 \text{ in.} \cdot \dfrac{2.54 \text{ cm}}{\text{in.}} = 22.86 \text{ cm}$

**47.** $26.5 \text{ m} \cdot \dfrac{1.09\text{yd}}{1 \text{ m}} \approx 28.89 \text{ yd}$

**49.** $19 \text{ L} \cdot \dfrac{0.264 \text{ gal}}{1 \text{ L}} \approx 5.02 \text{ gal}$

**51.** $32 \text{ lb} \cdot \dfrac{0.454 \text{ kg}}{1 \text{ lb}} \approx 14.53 \text{ kg}$

**53.** $F = 1.8C + 32 = 1.8(12) + 32$

$F = 53.6°F$

**55.** $C = \dfrac{5F - 160}{9}$

$C = \dfrac{5(68) - 160}{9} = 20°C$

**57.** $635 \text{ kg} \cdot \dfrac{2.2 \text{ lb}}{1 \text{ kg}} = 1397 \text{ lb}$

**59.** $67 \text{ mi} \cdot \dfrac{1.61 \text{ km}}{1 \text{ mi}} + 36 \text{ km} = 143.87 \text{ km}$

**61.** $15 \text{ gal} \cdot \dfrac{3.79 \text{ L}}{1 \text{ gal}} - 38 \text{ L} = 18.85 \text{ L}$

**63.** $F = 1.8(32) + 32 = 89.6°F$

$89.6°F - 80°F = 9.6°F$

**65.** $F = 1.8C + 32 = 1.8(19) + 32 = 66.2°F$

$F = 1.8C + 32 = 1.8(45) + 32 = 113°F$

**67.** $C = \dfrac{5F - 160}{9} = \dfrac{5(6188) - 160}{9}$

$C = 3420°C$

**69.** $0.768 \text{ oz} \cdot \dfrac{28.35 \text{ g}}{1 \text{ oz}} \approx 21.773 \text{ g}$

## Cumulative Review

**69.** $2^3 \times 6 - 4 + 3 = 8 \times 6 - 1 = 48 - 1 = 47$

**71.** $2^2 + 3^2 + 4^3 + 2 \times 7 = 4 + 9 + 64 + 14 = 91$

## 10.3 Exercises

**1.** Vertex: The point at which the two sides of an angle meet.

**3.** Obtuse angle: An angle whose measure is between 90° and 180°.

**5.** Adjacent angles: Two angles formed by intersecting lines that share a common side.

**7.** Line *u* is called a transversal since it intersects two or more lines at different points.

**9.** $\angle ABC$ is called a straight angle.

**11.** $\angle ABD$ is called an obtuse angle.

**13.** If the sum of two angles is $180°$ then these angles are called supplementary angles.

**15.** An angle whose measure is between $0°$ and $90°$ is called an acute angle.

**17.** $\angle a$ and $\angle c$ are called vertical angles.

**19.** $\angle d$ is equal to $\angle b$.

**21.** $\angle a + \angle b = 180°$.

**23.** Name the angles that are adjacent to $\angle c$: $\angle d$ and $\angle b$.

**25.** $\angle a$ and $\angle d$ are called alternate interior angles.

**27.** $\angle b$ is equal to the following two angles: $\angle c$ and $\angle f$.

**29.** $\angle f$ is equal to the following two angles: $\angle b$ and $\angle c$.

**31.** $\angle e$ and $\angle d$ are called corresponding angles.

**33.** $\angle b$ and $\angle f$ are called vertical angles.

**35.** $\angle XYZ$, $\angle ZYX$, or $\angle Y$.

**37.** $\angle TSV$ or $\angle VST$

**39.** (a) supplement of $\angle M = 180° - 29° = 151°$
(b) complement of $\angle M = 90° - 29° = 61°$

**41.** (a) supplement of $\angle X = 180° - 45° = 135°$
(b) complement of $\angle X = 90° - 45° = 45°$

**43.** $x + (x + 15°) = 90° \Rightarrow 2x = 90° - 15°$
$2x = 75° \Rightarrow x = 37.5°$

**45.** $n + (n - 5°) = 90° \Rightarrow 2n = 90° + 5°$
$2n = 95° \Rightarrow n = 47.5°$

**47.** $2m + m = 90° \Rightarrow 3m = 90° \Rightarrow m = 30°$

**49.** $\angle w$ and $\angle y$ are adjacent to $\angle x$.

**51.** $\angle w + \angle x = 180°$.

**53.** $\angle x = \angle z \Rightarrow \angle z = 65°, \angle w = 180° - \angle x$
$\angle w = 180° - 65° = 115° = \angle y$

**55.** $\angle w = 45° = \angle y,\ \angle x = 180° - \angle w$
$\angle x = 180° - 45° = 135° = \angle z$

**57.** $\angle p = \angle o = 43°,\ \angle o = \angle n = 43°$
$\angle n = \angle l = 43°,\ \angle m = 180° - 43° = 137°$

**59.** $\angle b = 180° - 45° = 135°,\ \angle a = \angle c = 45°$
$\angle c = \angle e = 45°,\ \angle d = 180° - 45° = 135°$

**61.** $\angle y = \angle v = \angle u = \angle r = 99°$
$\angle x = \angle w = \angle t = \angle s = 180° - 99° = 81°$

**63.** $2x + 2 = 5x - 10$
$3x = 12$
$x = 4$

**65.** $\angle x = \angle y = 33°$
$\angle w + \angle x = 90°,\ \angle w + \angle y = 90°$
$\angle w + 33° = 90°,\ \angle w = 57°$
$\angle z = 180° - \angle w = 180° - 57° = 123°$

### Cumulative Review

**67.** $V = LWH = 10(4)(8) = 320 \text{ m}^3$

**69.** $70 \text{ sq} \cdot \frac{\text{in.}}{14 \text{ sq}} + 2(2) \text{ in.} = 9 \text{ in.}$

$84 \text{ sq} \cdot \frac{\text{in.}}{14 \text{ sq}} + 2(2) \text{ in.} = 10 \text{ in.}$

9 in. by 10 in.

**How Am I Doing? Sections 10.1-10.3**

**1.** $21 \text{ ft} = 21 \text{ ft} \cdot \frac{1 \text{ yd}}{3 \text{ ft}} = 7 \text{ yd}$

**2.** $6 \text{ gal} \cdot \frac{4 \text{ qt}}{1 \text{ gal}} = 24 \text{ qt}$

**3.** $240 \text{ min} \cdot \frac{1 \text{ hr}}{60 \text{ min}} = 4 \text{ hr}$

**4.** $6.3 \text{ kg} = 6,300,000 \text{ mg}$

**5.** $34 \text{ mL} = 0.034 \text{ L} = 0.000034 \text{ kL}$

**6.** $63 \text{ ft} \cdot \frac{1 \text{ yd}}{3 \text{ ft}} \cdot \frac{\$4.00}{\text{yd}} = \$84$

**7.** $10 \text{ in.} \cdot \frac{2.54 \text{ cm}}{\text{in.}} = 25.4 \text{ cm}$

**8.** $22 \text{ lb} \cdot \frac{0.454 \text{ kg}}{1 \text{ lb}} = 9.99 \text{ kg}$

**9.** $3.5 \text{ L} \cdot \frac{1.06 \text{ qt}}{\text{L}} = 3.71 \text{ qt}$

**10.** $\frac{95 \text{ km}}{\text{hr}} \cdot \frac{0.62 \text{ mi}}{\text{km}} = \frac{58.9 \text{ mi}}{\text{hr}}$

**11.** $(1.8 \cdot 35°\text{C} + 32) - 89°\text{F} = 6°\text{F}$

**12.** $\angle c = 45°$

**13.** $\angle y = 32°$

**14.** (a) supplement of $\angle a = 180° - 39° = 141°$
(b) compliment of $\angle a = 90° - 39° = 51°$

**15.**
$$x + (x + 40°) = 90°$$
$$2x + 40° = 90°$$
$$2x = 50°$$
$$x = 25°$$

**16.**
$$\angle z = \angle x = 55°$$
$$\angle w + \angle x = 180°$$
$$\angle w + 55° = 180°$$
$$\angle w + 55° = 180°$$
$$\angle w = 180° - 55° = 125°$$
$$\angle y = \angle w = 125°$$

**17.** $\angle n = \angle o = \angle p = 40°$
$\angle l = \angle m = 180° - 40° = 140°$

**10.4 Exercises**

**1.** (a) $\sqrt{81}$: What is the square root of 81?
(b) $n \cdot n = 81$: What number multiplied by itself equals 81?
(c) $n^2 = 81$: What number squared equals 81?

**3.** $6^2 = 36,\ 7^2 = 49 \Rightarrow 44$ is not a perfect square.

**5.** $6^2 = 36 \Rightarrow 36$ is a perfect square.

**7.** $9^2 = 81 \Rightarrow 81$ is a perfect square.

**9.** $\left(\frac{1}{2}\right)^2 = \frac{1}{4},\ \left(\frac{1}{3}\right)^2 = \frac{1}{9} \Rightarrow \frac{1}{5}$ is not a perfect square.

**11.** (a) $n^2 = 64 \Leftrightarrow n = \sqrt{64}$

(b) $\sqrt{64}$ is the positive square root of 64.

(c) $n \cdot n = 64 \Leftrightarrow n = \sqrt{64}$

**13.** $\sqrt{49} = 7$

**15.** $\sqrt{64} = 8$

**17.** $\sqrt{144} = 12$

**19.** $\sqrt{225} - \sqrt{16} = 15 - 4 = 11$

**21.** $\sqrt{9} + \sqrt{49} = 3 + 7 = 10$

**23.** $\sqrt{\frac{25}{49}} = \frac{5}{7}$

**25.** $\sqrt{100} - \sqrt{25} = 10 - 5 = 5$

**27.** $\sqrt{121} + \sqrt{81} = 11 + 9 = 20$

**29.** $\sqrt{\frac{9}{81}} = \frac{3}{9} = \frac{1}{3}$

**31.** $\sqrt{\frac{36}{49}} = \frac{6}{7}$

**33.** $A = s^2 = 121 \Rightarrow s = 11$ ft

**35.** $A = s^2 = 144 \Rightarrow s = 12$ in.

**37.** $\sqrt{31} \approx 5.568$

**39.** $\sqrt{69} \approx 8.307$

**41.** $\sqrt{80} \approx 8.944$

**43.** $\sqrt{90} \approx 9.487$

**45.** $c^2 = 3^2 + 4^2 = 9 + 16 = 25 \Rightarrow c = 5$ in.

**47.** $c^2 = 8^2 + 3^2 = 64 + 9 = 73$

$c = \sqrt{73} \approx 8.544$ yd

**49.** $16^2 = a^2 + 5^2 \Rightarrow a^2 = 256 - 25 = 231$

$a = \sqrt{231} \approx 15.199$ ft

**51.** $13^2 = 8^2 + b^2 \Rightarrow b^2 = 169 - 64 = 105$

$b = \sqrt{105} \approx 10.247$ km

**53.** $c^2 = 11^2 + 3^2 = 121 + 9 = 130$

$c = \sqrt{130} \approx 11.402$ m

**55.** $c^2 = 5^2 + 5^2 = 25 + 25 = 50$

$c = \sqrt{50} \approx 7.071$ m

**57.** $5^2 = 4^2 + x^2 \Rightarrow x^2 = 5^2 - 4^2 = 25 - 16 = 9$

$x^2 = 9 \Rightarrow x = 3,\ A = 4^2 + \frac{1}{2} \cdot (3)(4)$

$A = 16 + 6 = 22$ in.$^2$

**59.** $13^2 = 12^2 + x^2 \Rightarrow x^2 = 169 - 144 = 25$

$x = 5,\ A = 4(12) + \frac{1}{2} \cdot (5)(12)$

$A = 48 + 30 = 78$ ft$^2$

**61.** $c^2 = 8^2 + 15^2 = 64 + 225 = 289$

$c = \sqrt{289} = 17$ ft

**63.** $c^2 = 3^2 + 4^2 = 9 + 16 = 25$

$c = \sqrt{25} = 5$ mi

**65.** $20^2 = 18^2 + x^2 \Rightarrow x^2 = 400 - 324 = 76$

$x = \sqrt{76} \approx 8.7$ ft

**67.** $A = 2(15)(20) + 2(15)(30) + 2 \cdot \frac{1}{2} \cdot (20)(12)$

$A = 1740 \text{ ft}^2$

**69.** $C = 2 \cdot \frac{1}{2} \cdot (18)(13) \cdot 90 = \$21,060$

**Cumulative Review**

**71.** $P = a + b + c = 3 + 4 + 5 = 12 \text{ ft}$

**73.** $A = LW = 8(4) = 32 \text{ cm}^2$

**10.5 Exercises**

**1.** The distance around a circle is called the <u>circumference</u>.

**3.** The diameter is two times the <u>radius</u> of the circle.

**5.** $d = 2r = 45 \Rightarrow r = 22.5 \text{ yd}$

**7.** $d = 2r = 3.8 \Rightarrow r = 1.9 \text{ cm}$

**9.** $C = \pi d \approx 3.14(32) \approx 100.5 \text{ cm}$

**11.** $C = 2\pi r \approx 2(3.14)(11) \approx 69.1 \text{ in.}$

**13.** distance $= C = \pi d \approx 3.14(28) \approx 87.9 \text{ in.}$

**15.** distance $= 5C = 5\pi d \approx 5 \cdot 3.14(24)$
distance $\approx 376.8$ in.

**17.** $A = \pi r^2 \approx 3.14(7)^2 \approx 153.9 \text{ yd}^2$

**19.** $A = \pi r^2 \approx 3.14\left(\frac{44}{2}\right)^2 \approx 1519.8 \text{ cm}^2$

**21.** $A = \pi r^2 \approx 3.14(12)^2 \approx 452.2 \text{ ft}^2$

**23.** $A = \pi r^2 \approx 3.14\left(\frac{90}{2}\right)^2 = 6358.5 \text{ mi}^2$

**25.** $C = \pi d \approx 3.14(2) = 6.28 \text{ ft}$

**27.** distance $= 35\pi d \approx 35(3.14)(14 \cdot 2)$
distance $\approx 3077.2$ in.
distance $\approx 3077.2 \text{ in.} \cdot \frac{1 \text{ ft}}{12 \text{ in.}} \approx 256.43 \text{ ft}$

**29.** distance $= 20,096 = n \cdot 2\pi r$
$20,096 \approx n \cdot 2(3.14)(16)$
$n \approx 200$ rev

**31.** $A = \pi r^2 \approx 3.14\left(\frac{64}{2}\right)^2 = 3215.36 \text{ in.}^2$

**33.** $A = \pi r^2 \approx 3.14\left(\frac{6}{2}\right)^2 = 28.26 \text{ ft}^2$

$\text{cost} = 28.26 \text{ ft}^2 \cdot \frac{1 \text{ yd}^2}{9 \text{ ft}^2} \cdot \frac{\$72}{\text{yd}^2} = \$226.08$

**35.** $A = LW + \pi r^2 \approx 120(40) + 3.14\left(\frac{40}{2}\right)^2$

$A \approx 6056 \text{ yd}^2$, $C = 6056(0.20) = \$1211.20$

**37.** (a) $\frac{\$6}{8 \text{ slices}} = \$0.75$ per slice

$\frac{1}{8} \cdot \pi r^2 \approx \frac{1}{8} \cdot 3.14\left(\frac{15}{2}\right)^2 \approx 22.1 \text{ in.}^2$

(b) $\frac{\$4}{6 \text{ slices}} = \$0.67$ per slice

$\frac{1}{6} \cdot \pi r^2 \approx \frac{1}{6} \cdot 3.14\left(\frac{12}{2}\right)^2 \approx 18.8 \text{ in.}^2$

(c) 12 in.: $\frac{\$0.67}{18.8 \text{ in.}^2} \approx \$0.036$ per in.$^2$

15 in.: $\frac{\$0.75}{22.1 \text{ in.}^2} \approx \$0.034$ per in.$^2$

The 15 in. pizza is the better buy.

**39.** $C = \pi d = \pi(0.223) \approx 0.70058$ m

**Cumulative Review**

**41.** $0.16(87) = 13.92$

**43.** $V = LWH = 11(5)(6) = 330 \text{ in.}^3$

**Putting Your Skills to Work, Problems for Individual Investigation**

**1.** Patio: $40(15) = 600 \text{ ft}^2$

Planter: $50(3) + 3(57) + 3(47) = 462 \text{ ft}^2$

Lawn:

$50(60) - 15(40) - 600 - 462 = 1338 \text{ ft}^2$

**2.** $600 \text{ ft}^2 \cdot 4 \text{ in.} \cdot \dfrac{1 \text{ ft}}{12 \text{ in.}} = 200 \text{ ft}^3$

**3.** $(50-3)(2) + (60-6) = 148 \text{ ft} \cdot \dfrac{12 \text{ in.}}{1 \text{ ft}}$
$= 1776 \text{ in.}$

$1776 \text{ in.} \cdot \dfrac{1 \text{ brick}}{8 \text{ in.}} \cdot 4 = 888 \text{ bricks}$

**4.** $462 \text{ ft}^2 \cdot 12 \text{ in.} \cdot \dfrac{1 \text{ ft}}{12 \text{ in.}} = 462 \text{ ft}^3$

**5.** $2(10) + 2(50) + 60 = 180$ ft of fence

**Putting Your Skills to Work, Problems for Group Investigation and Cooperative Study**

**1.** $1 \text{ yd}^3 = 1 \text{ yd}^3 \cdot \dfrac{3^3 \text{ ft}^3}{\text{yd}^3} = 27 \text{ ft}^3$

**2.** Cement:

$200 \text{ ft}^3 \cdot \dfrac{1 \text{ yd}^3}{27 \text{ ft}^3} \approx 7.4 \text{ yd}^3 \to 8 \text{ yd}^3$

Planter:

$462 \text{ ft}^3 \cdot \dfrac{1 \text{ yd}^3}{27 \text{ ft}^3} \approx 17.1 \text{ yd}^3 \to 18 \text{ yd}^3$

**3.** $A = \pi r^2 \approx 3.14\left(\dfrac{8}{2} + 2\right)^2 = 113.04 \text{ ft}^2$

$1338 - 113.04 = 1224.96 \text{ ft}^2$

**4.** (a) Answers may vary.
(b) Answers may vary.

**10.6 Exercises**

**1.** $V = LWH$: box

**3.** $V = \dfrac{4\pi r^3}{3}$: sphere

**5.** $V = \dfrac{Bh}{3}$: pyramid

**7.** $V = \pi r^2 h \approx 3.14(2)^2(7) \approx 87.9 \text{ m}^3$

**9.** $V = \dfrac{4\pi r^3}{3} \approx \dfrac{4(3.14)(4)^3}{3} \approx 267.9 \text{ m}^3$

**11.** $V = \pi r^2 h = 3.14(8.3)^2(14.4)$

$V = 3114.9 \text{ cm}^3$

**13.** $V = \dfrac{4}{3}\pi r^3 = \dfrac{4}{3} \cdot 3.14(3.25)^3$

$V = 143.7 \text{ cm}^3$

**15.** $V = \dfrac{1}{2} \cdot \dfrac{4\pi r^3}{3} \approx \dfrac{1}{2} \cdot \dfrac{4(3.14)(7)^3}{3}$

$V \approx 718.0 \text{ m}^3$

**17.** $V = \dfrac{\pi r^2 h}{3} \approx \dfrac{3.14(9)^2(12)}{3} \approx 1017.4 \text{ cm}^3$

**19.** $V = \frac{\pi r^2 h}{3} \approx \frac{3.14(5)^2(10)}{3} \approx 261.7 \text{ ft}^3$

**21.** $V = \frac{Bh}{3} = \frac{3^2(7)}{3} = 21 \text{ m}^3$

**23.** $V = \frac{Bh}{3} = \frac{6(12)(5)}{3} = 120 \text{ m}^3$

**25.** $V = \pi(R^2 - r^2)h \approx 3.14(5^2 - 3^2)(20)$
$V \approx 1004.8 \text{ in.}^3$

**27.** $V = \pi(R^2 - r^2)h = (3.14)(8^2 - 5^2)(30)$
$V = 3673.8 \text{ in.}^3$

**29.** $\frac{\$0.09}{\text{in.}^3} \cdot \left( \frac{3.14(1)^2(6)}{3} + \frac{1}{2} \cdot \frac{4(3.14)(1)^2}{3} \right)$
$= \$0.75$

**31.** $V = LWH + \pi r^2 h \approx 4(3)(2) + 3.14(1)^2(2)$
$V \approx 30.28 \text{ ft}^3$

**33.** $\text{cost} = \frac{\$4}{\text{cm}^3} \cdot \frac{3.14(5)^2(9)}{3} = \$942$

**35.** $S = 4\pi r^2 \approx 4(3.14)(7)^2 = 615.44 \text{ in.}^2$

**37.** $V = \frac{4\pi r^3}{3} \approx \frac{4(3.14)(5.21)^3}{3} \approx 592.1 \text{ m}^3$

**39.** $V = \frac{Bh}{3} = \frac{6.22(5.01)(9.212)}{3} \approx 95.7 \text{ ft}^3$

**Cumulative Review**

**41.** $\frac{21}{40} = \frac{x}{120} \Rightarrow 40x = 21(120) = 2520$
$x = \frac{2520}{40} = 63$

**43.** $2\frac{1}{4} \cdot 3\frac{3}{4} = \frac{9}{4} \cdot \frac{15}{4} = \frac{135}{16} = 8\frac{7}{16}$

**10.7 Exercises**

**1.** The corresponding angles of similar triangles are equal.

**3.** The perimeters of similar triangles have the same ratios as the corresponding sides.

**5.** $\frac{3}{2} = \frac{12}{n} \Rightarrow 3n = 24 \Rightarrow n = \frac{24}{3} = 8 \text{ m}$

**7.** $\frac{9}{5} = \frac{27}{n} \Rightarrow 9n = 135 \Rightarrow n = \frac{135}{9} = 15 \text{ cm}$

**9.** $\frac{7}{2} = \frac{n}{5} \Rightarrow 2n = 7(5) = 35 \Rightarrow n = 17.5 \text{ cm}$

**11.** $\frac{18}{5} = \frac{8}{n} \Rightarrow 18n = 40 \Rightarrow n = 2.2 \text{ in.}$

**13.** $\frac{P}{22} = \frac{10+6+9}{10} \Rightarrow 10P = 22(25) = 550$
$P = 55 \text{ in.}$

**15.** $\frac{P}{10} = \frac{20+16+18}{20} \Rightarrow 20P = 10(54) = 540$
$P = 27 \text{ cm}$

**17.** $\frac{9}{5} = \frac{90}{n} \Rightarrow 9n = 450 \Rightarrow n = 50 \text{ cm}$

**19.** $\frac{n}{24} = \frac{6}{4} \Rightarrow 4n = 144 \Rightarrow n = 36 \text{ ft}$

**21.** $\frac{5.5}{6.5} = \frac{n}{96} \Rightarrow 6.5n = 528 \Rightarrow n \approx 81 \text{ ft}$

**23.** $\frac{20}{42}=\frac{n}{8}\Rightarrow 42n=160\Rightarrow n\approx 3.8$ km

**25.** $\frac{n}{14}=\frac{15}{9}\Rightarrow 9n=210\Rightarrow n\approx 23.3$ cm

**27.** $\frac{9}{12}=\frac{n}{19}\Rightarrow 12n=171\Rightarrow n=14.25$ ft

**29.** $\frac{26}{3^2}=\frac{A}{2^2}\Rightarrow 9A=104\Rightarrow A\approx 11.6\text{ yd}^2$

**Cumulative Review**

**31.** $2\cdot 3^2+4-2\cdot 5$
$=2\cdot 9+4-10=18-6=12$

**33.** $(5)(9)-(21+3)\div 8$
$=45-24\div 8=45-3=42$

**Chapter 10 Review Problems**

**1.** right angle – an angle that measures $90°$

**2.** supplementary angles – two angles whose sum is $180°$

**3.** complementary angles – two angles whose sum is $90°$

**4.** vertical angles – two angles that are opposite each other

**5.** adjacent angles – two angles formed by intersecting lines that share a common side

**6.** alternating interior angles – two angles that are on opposite sides of the transversal and between the two parallel lines.

**7.** corresponding angles – two angles that are on the same side of the transversal and are both above or below the other two lines

**8.** right triangle – a triangle with a $90°$ angle

**9.** Pythagorean Theorem – in a right triangle the square of the longest side is equal to the sum of the squares of the other two sides

**10.** radius – a line segment from the center to a point on the circle

**11.** diameter – a line segment across the circle that passes through the center

**12.** circumference – the distance around a circle

**13.** area of a triangle: $A=\frac{1}{2}bh$

**14.** circumference of a circle: $C=\pi d=2\pi r$

**15.** area of a circle: $A=\pi r^2$

**16.** volume of a cylinder: $V=\pi r^2 h$

**17.** volume of a sphere: $V=\frac{4}{3}\pi r^3$

**18.** volume of a cone: $V=\frac{1}{3}\pi r^2 h$

**19.** volume of a pyramid: $V=\frac{1}{3}Bh$

**20.** $27\text{ ft}\cdot\frac{1\text{ yd}}{3\text{ ft}}=9\text{ yd}$

**21.** $2160 \text{ sec} \cdot \frac{1 \text{ min}}{60 \text{ sec}} = 36 \text{ min}$

**22.** $90 \text{ in.} \cdot \frac{1 \text{ ft}}{12 \text{ in.}} = 7.5 \text{ ft}$

**23.** $15,840 \text{ ft} \cdot \frac{1 \text{ mi}}{5280 \text{ ft}} = 3 \text{ mi}$

**24.** $4 \text{ tons} \cdot \frac{2000 \text{ lb}}{1 \text{ ton}} = 8000 \text{ lb}$

**25.** $15 \text{ gal} \cdot \frac{4 \text{ qt}}{1 \text{ gal}} = 60 \text{ qt}$

**26.** $92 \text{ oz} \cdot \frac{1 \text{ lb}}{16 \text{ oz}} = 5.75 \text{ lb}$

**27.** $31 \text{ pt} \cdot \frac{1 \text{ qt}}{2 \text{ pt}} = 15.5 \text{ qt}$

**28.** 59 mL = 0.059 L

**29.** 56 cm = 560 mm

**30.** 2598 mm = 259.8 cm

**31.** 778 mg = 0.778 g

**32.** 9.2 m = 920 cm

**33.** 7 km = 7000 m

**34.** 17 kL = 17,000 L

**35.** 473 m = 0.473 km

**36.** 196 kg = 196,000 g

**37.** 721 kg = 721,000 g

**38.** $\frac{4 \text{ L}}{24 \text{ jars}} = \frac{4000 \text{ mL}}{24 \text{ jars}} \approx 166.67 \ \frac{\text{mL}}{\text{jar}}$

**39.** (a) $P = 7\frac{2}{3} + 4\frac{1}{3} + 5 = 17 \text{ ft}$

(b) $P = 17 \text{ ft} \cdot \frac{12 \text{ in.}}{1 \text{ ft}} = 204 \text{ in.}$

**40.** $42 \text{ kg} \cdot \frac{2.2 \text{ lb}}{\text{kg}} = 92.4 \text{ lb}$

**41.** $20 \text{ lb} \cdot \frac{0.454 \text{ kg}}{\text{lb}} = 9.08 \text{ kg}$

**42.** $15 \text{ ft} \cdot \frac{0.305\text{m}}{\text{ft}} \approx 4.58 \text{ m}$

**43.** $1.8 \text{ ft} \cdot \frac{12 \text{ in.}}{\text{ft}} \cdot \frac{2.54 \text{ cm}}{\text{in.}} \approx 54.86 \text{ cm}$

**44.** $13 \text{ oz} \cdot \frac{28.35 \text{ g}}{\text{oz}} = 368.55 \text{ g}$

**45.** $F = 1.8C + 32 = 1.8(15) + 32 = 59°\text{F}$

**46.** $14 \text{ cm} \cdot \frac{0.394 \text{ in.}}{\text{cm}} \approx 5.52 \text{ in.}$

**47.** $C = \frac{5F - 160}{9} = \frac{5(32) - 160}{9} = 0°\text{C}$

**48.** $200 \text{ mi} - 90\frac{\text{km}}{\text{hr}} \cdot 3 \text{ hr} \cdot \frac{0.62 \text{ mi}}{\text{km}} = 32.6 \text{ mi}$

**49.** $\frac{\$0.14}{\text{oz}} \cdot 450 \text{ g} \cdot \frac{0.0353 \text{ oz}}{\text{g}} = \$2.22$

**50.** $\angle MNO$, $\angle ONM$

**51.** $\angle PNO$, $\angle$ONP

**52.** (a) $180° - 39° = 141°$
(b) $90° - 39° = 51°$

**53.** $m + (m + 10°) = 90° \Rightarrow 2m = 80°$
$m = \angle ABD = 40°$

**54.** $\angle w, \angle y$

**55.** $\angle x + \angle y = 180°$

**56.** $\angle z = \angle x = 65°$
$\angle w = \angle y = 180° - 65° = 115°$

**57.** $\angle w = \angle v = \angle u = \angle s = 45°$
$\angle t = 180° - 45° = 135°$

**58.** $\sqrt{64} = 8$

**59.** $\sqrt{\frac{64}{81}} = \frac{8}{9}$

**60.** $\sqrt{45} \approx 6.708$

**61.** $\sqrt{5} \approx 2.236$

**62.** $\sqrt{64} - \sqrt{49} = 8 - 7 = 1$

**63.** $\sqrt{144} + \sqrt{9} = 12 + 3 = 15$

**64.** $A = 5.1^2 + \frac{1}{2} \cdot (9.6)(5.1) \approx 50.5 \text{ m}^2$

**65.** $13^2 = 12^2 + a^2 \Rightarrow a^2 = 169 - 144 = 25$
$a = 5 \text{ yd}$

**66.** $7^2 = 6^2 + a^2 \Rightarrow a^2 = 49 - 36 = 13$
$a = \sqrt{13} \approx 3.61 \text{ ft}$

**67.** $C = \pi d \approx 3.14(12) \approx 37.7 \text{ in.}$

**68.** $C = 2\pi r \approx 2(3.14)(7) \approx 44.0 \text{ in.}$

**69.** $A = \pi r^2 \approx 3.14(6)^2 = 113.04 \text{ m}^2$

**70.** $A = \pi r^2 \approx 3.14\left(\frac{16}{2}\right)^2 = 200.96 \text{ ft}^2$

**71.** $3.14(5 \text{ ft})^2 \cdot \frac{1 \text{ yd}^2}{9 \text{ ft}^2} \cdot \frac{\$30}{\text{yd}^2} = \$261.67$

**72.** $V = \frac{4\pi r^3}{3} \approx \frac{4(3.14)(1.2)^3}{3} \approx 7.2 \text{ ft}^3$

**73.** (a) $V = LWH - \pi r^2 h$
$V \approx 4(3)(7) - 3.14(1)^2(7) \approx 62.02 \text{ m}^3$
(b) $\text{cost} = 62.02 \text{ m}^3 \cdot \frac{\$1.20}{\text{m}^3} = \$74.42$

**74.** $V = \pi r^2 h + \frac{1}{2} \cdot \frac{4\pi r^3}{3}$
$V \approx 3.14(2)^2(9) + \frac{1}{2} \cdot \frac{4(3.14)(2)^3}{3}$
$V \approx 129.787 \text{ cm}^3$

**75.** $V = \frac{Bh}{3} = \frac{16(18)(18)}{3} = 1728 \text{ m}^3$

**76.** $V = \frac{\pi r^2 h}{3} \approx \frac{3.14(17)^2(30)}{3} = 9074.6 \text{ yd}^3$

**77.** $\frac{3}{2} = \frac{45}{n} \Rightarrow 3n = 90 \Rightarrow n = 30 \text{ m}$

**78.** $\frac{18 + 5 + 26 + 5}{18} = \frac{P}{108} \Rightarrow 18P = 5832$
$P = 324 \text{ cm}$

## How Am I Doing? Chapter 10 Test

**1.** $145 \text{ oz} = 144 \text{ oz} \cdot \frac{1 \text{ lb}}{16 \text{ oz}} + 1 \text{ oz} = 9 \text{ lb } 1 \text{ oz}$

**2.** $162 \text{ g} = 0.162 \text{ kg}$

**3.** (a) $2 \text{ kg} = 2000 \text{ g}$
(b) $2 \text{ kg} \cdot \frac{2.2 \text{ lb}}{\text{kg}} = 4.4 \text{ lb}$

**4.** $F = 1.8C + 32 = 1.8(200) + 32 = 392°\text{F}$

**5.** $14 \text{ km} + 4 \text{ mi} \cdot \frac{1.61 \text{ km}}{\text{mi}} = 20.44 \text{ km}$

**6.** (a) $180° - 41° = 139°$
(b) $90° - 41° = 49°$

**7.** $m + (m + 8°) = 90° \Rightarrow 2m = 82°$
$m = 41°$

**8.** (a) $\angle a = \angle c = 70°$
(b) $\angle d = 180° - 70° = 110°$

**9.** $\angle w = \angle u = \angle t = 43°$
$\angle s = \angle v = 180° - 43° = 137°$

**10.** $81 = 9^2$, yes

**11.** $4^2 = 16,\ 5^2 = 25$
24 is not a perfect square

**12.** $\frac{1}{2^2} = \frac{1}{4},\ \frac{1}{3^2} = \frac{1}{9}$
$\frac{1}{8}$ is not a perfect square

**13.** $A = s^2 = 64 \Rightarrow s = 8 \text{ ft}$

**14.** $\sqrt{49} = 7$

**15.** $\sqrt{\frac{9}{16}} = \frac{3}{4}$

**16.** $\sqrt{6} \approx 2.45$

**17.** $\sqrt{25} + \sqrt{36} = 5 + 6 = 11$

**18.** $A = \frac{1}{2}bh = \frac{1}{2}(10)(12) = 60 \text{ cm}^2$

**19.** $A = 30(20) + \frac{1}{2}(20)(10) = 700 \text{ ft}^2$

**20.** $7^2 = 3^2 + a^2 \Rightarrow a^2 = 40 \Rightarrow a = \sqrt{40}$
$a \approx 6.32 \text{ cm}$

**21.** $c^2 = 14^2 + 11^2 = 317 \Rightarrow c = \sqrt{317}$
$c \approx 17.8 \text{ ft}$

**22.** $C = \pi d \approx 3.14(5.15) = 16.171 \text{ m}$

**23.** $2\pi R = n\pi r \Rightarrow 2(2.1) = n(0.75)$
$n = 5.6 \text{ rev}$

**24.** $A = \pi r^2 \approx 3.14(1.2)^2 \approx 4.5 \text{ cm}^2$

**25.** $\frac{\$20}{\text{yd}^2} \cdot 3.14(3 \text{ ft})^2 \cdot \frac{\text{yd}^2}{9 \text{ ft}^2} = \$62.80$

**26.** $V = \pi r^2 h \approx 3.14(3.4)^2(7.1) \approx 257.72 \text{ cm}^3$

**27.** $V = \frac{4\pi r^3}{3} \approx \frac{4(3.14)(8)^3}{3} \approx 2143.57 \text{ in.}^3$

**28.** $V = \frac{\pi r^2 h}{3} \approx \frac{3.14(8)^2(12)}{3} = 803.84 \text{ cm}^3$

**29.** $V = \frac{Bh}{3} = \frac{10(7)(12)}{3} = 280 \text{ m}^3$

**30.** $\frac{4}{13} = \frac{18}{n} \Rightarrow 4n = 234 \Rightarrow n = 58.5 \text{ in.}$

**31.** $\frac{h}{2.5} = \frac{20}{2} \Rightarrow 2h = 50 \Rightarrow h = 25 \text{ ft}$

**32.** $\frac{1}{4.5} = \frac{W}{36} \Rightarrow 4.5W = 36 \Rightarrow W = 8 \text{ cm}$

## Cumulative Test for Chapters 1-10

**1.** $0 > -5$

**2.** $0.5 > 0.15$

**3.** $-8 > -18$

**4.** $8 < 116$

**5.** $|-7| = 7,\ |-9| = 9 \Rightarrow |-9| > |-7|$

**6.** $|-6| = 6,\ |2| = 2 \Rightarrow |-6| > |2|$

**7.** (a) $2(9) + 2(12) = 42 \text{ ft}$
(b) $2.5(42) = \$105$
(c) $9(12) = 108 \text{ ft}^2$

**8.** $12 - 5^2 + 3 = 12 - 25 + 3 = -13 + 3 = -10$

**9.** $4(8 - 12) = 4(-4) = -16$

**10.** $-15(-3) = 45$

**11.** $\frac{1}{3}\left(\frac{2}{5}\right) \cdot \frac{2}{2} + \frac{1}{2} \cdot \frac{15}{15} = \frac{4 + 15}{30} = \frac{19}{30}$

**12.** $\frac{-3}{4} \div \frac{7}{8} = -\frac{3}{4} \cdot \frac{8}{7} = -\frac{6}{7}$

**13.** $2x + 3x^2 - 8 + x - 17 = 3x^2 + 3x - 25$

**14.** $x - 20 = 18 \Rightarrow x = 18 + 20 = 38$
check: $38 - 20 \stackrel{?}{=} 18,\ 18 = 18$

**15.** $x + 3 = -12 \Rightarrow x = -12 - 3 = -15$
check: $-15 + 3 \stackrel{?}{=} -12,\ -12 = -12$

**16.** $3 - 8 + x = 14 - 9 \Rightarrow -5 + x = 5 \Rightarrow x = 10$
check: $3 - 8 + 10 \stackrel{?}{=} 14 - 9,\ 5 = 5$

**17.** $\frac{x}{6} = 4 \Rightarrow x = 6(4) = 24$
check: $\frac{24}{6} \stackrel{?}{=} 4,\ 4 = 4$

**18.** $5x - 9(2) = 48 \Rightarrow 5x = 48 + 18 = 66$
$x = \frac{66}{5} = 13.2$
check: $5(13.2) - 9(2) \stackrel{?}{=} 48,\ 48 = 48$

**19.** $4(x + 3) = 4x + 4(3) = 4x + 12$

**20.** $(x + 2)(x - 1) = x^2 - x + 2x - 2$
$= x^2 + x - 2$

**21.** $(4x + 3y - 9) + (-3x + 7y + 6)$
$= 4x + 3y - 9 - 3x + 7y + 6$
$= x + 10y - 3$

**22.** $0.83(250) = \$207.50$

**23.** $\frac{26,275 - 23,450}{23,450} \approx 0.120469 \to 12.05\%$

**24.** $\frac{76+85+73+75+81}{5}=78$

**25.** 73, 75, 76, 81, 85

median = 76

**26.** (a) $y=2x+1$

$y=2(-1)+1=-1 \Rightarrow (-1,-1)$

$y=2(0)+1=1 \Rightarrow (0,1)$

$y=2(1)+1=3 \Rightarrow (1,3)$

(b)

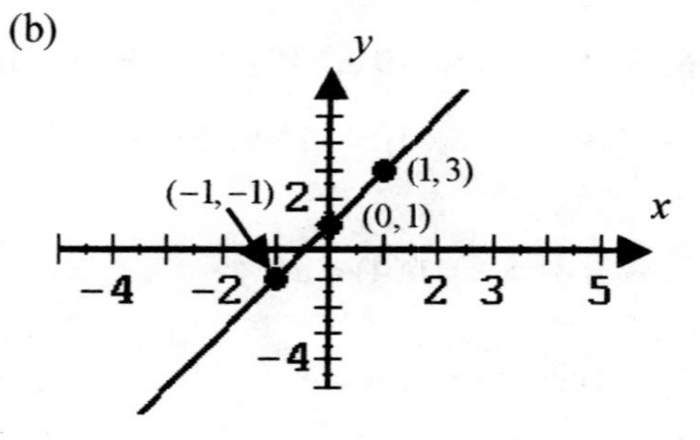


**27.** (a) $96 \text{ in.} \cdot \frac{1 \text{ ft}}{12 \text{ in.}} = 8 \text{ ft}$

(b) $2.25 \text{ lb} \cdot \frac{16 \text{ oz}}{\text{lb}} = 36 \text{ oz}$

**28.** (a) 57 cm = 570 mm

(b) 4216 mL = 0.004216 kL

**29.** (a) $280 \text{ gal} \cdot \frac{3.79 \text{ L}}{\text{gal}} = 1061.2 \text{ L}$

(b) $18 \text{ m} \cdot \frac{1.09 \text{ yd}}{\text{m}} = 19.62 \text{ yd}$

(c) $186 \text{ g} \cdot \frac{0.0353 \text{ oz}}{\text{g}} \approx 6.57 \text{ oz}$

**30.** $\angle ZXY$, $\angle YXZ$

**31.** $\angle e = \angle g = \angle i = 35°$

$\angle f = \angle h = 180° - 35° = 145°$

**32.** supplement of $\angle B = 180° - 42° = 138°$

complement of $\angle B = 90° - 42° = 48°$

**33.** $A = 4^2 + \frac{1}{2} \cdot 16 \cdot 4 = 48 \text{ in.}^2$

**34.** $c^2 = 3^2 + 4^2 = 9 + 16 = 25$

$c = 5$ ft

**35.** $\sqrt{144} = 12$

**36.** $\sqrt{\frac{81}{100}} = \frac{9}{10}$

**37.** $\sqrt{10} \approx 3.2$

**38.** $c^2 = 6^2 + 8^2 = 36 + 64 = 100 \Rightarrow c = 10$ m

**39.** $C = \pi d \approx 3.14(14) = 43.96$ in.

**40.** $V = \pi r^2 h \approx 3.14(2.7)^2(6.3) \approx 144.21 \text{ cm}^3$

**41.** $V = \frac{Bh}{3} = \frac{14(10)(12)}{3} = 560 \text{ in.}^3$

**42.** $\frac{h}{20} = \frac{15}{10} \Rightarrow 10h = 300 \Rightarrow h = 30$ ft

### Practice Final Examination

**1.** 7,543,876

= 7,500,000 nearest hundred thousand

**2.** A number decreased by 5: $x - 5$

**3.** $(2+5+x)+4 = (7+x)+4 = (x+7)+4$

$= x + (7+4) = x + 11$

**4.** $x + y\big|_{x=9, y=12} = 9 + 12 = 21$

**5.** $6+5+15+4+(15+6)+(5+4) = 60$ ft

**6.** 
$$\begin{array}{r} 751 \\ -482 \\ \hline 269 \end{array}$$

check: $269+482\overset{?}{=}751,\ 751=751$

**7.** 
$$\begin{array}{r} 5282 \\ \times 806 \\ \hline 31692 \\ 42256\phantom{00} \\ \hline 4257292 \end{array} \rightarrow 4,257,292$$

**8.** $812,869 \div 743 = 1094 \text{ R}27$

$$\begin{array}{r} 1094 \\ 743\overline{)812869} \\ 743\phantom{000} \\ \hline 6986\phantom{00} \\ 6687\phantom{00} \\ \hline 2999\phantom{0} \\ 2972\phantom{0} \\ \hline 27 \end{array}$$

**9.** $7+15\div 3-3^2=7+5-9=12-9=3$

**10.** $y+4y+3y=16 \Rightarrow 8y=16 \Rightarrow y=2$

**11.** $30+30+70+190=\$320$

**12.** $-5<-3$

**13.** $-|-13|=-(-(-13))=-(13)=-13$

**14.** $7+(-3)+9+(-4)=4+5=9$

**15.** $-9-7+8=-16+8=-8$

**16.** $(-6)(2)(-1)(5)(-3)=(-12)(-5)(-3)$
$=60(-3)=-180$

**17.** $64\div(-8)=-8$

**18.** $-9+5y+7-3y=5y-3y-9+7$
$=2y-2$

**19.** $a^2-b\big|_{a=-4,b=7}=(-4)^2-7=16-7=9$

**20.** $-3(a-2)=-3a-(-3)(2)=-3a+6$

**21.** $5(3-7)=x-3 \Rightarrow 5(-4)=x-3$
$x-3=-20 \Rightarrow x=-20+3=-17$

check: $5(3-7)\overset{?}{=}-17-3,\ -20=-20$

**22.** $6(3x)=54 \Rightarrow 18x=54 \Rightarrow x=3$

check: $6(3(3))\overset{?}{=}54,\ 54=54$

**23.** (a) $L=5W$

(b) $40=5W \Rightarrow W=\frac{40}{5}=8 \text{ ft}$

**24.** $21(10)+(21-14)(9)+7(11)=350 \text{ in.}^2$

**25.** $A=bh=b(9)=108 \Rightarrow b=12 \text{ m}$

**26.** $V=10W(5)=200 \Rightarrow W=\frac{200}{50}=4 \text{ ft}$

**27.** (a) $(21 \text{ ft})(15 \text{ ft})\cdot\frac{1 \text{ yd}^2}{9 \text{ ft}^2}=35 \text{ yd}^2$

(b) $35 \text{ yd}^2\cdot\frac{\$18}{\text{yd}^2}=\$630$

**28.** $(5y)(z^6)(5y^4)(3z^2)=5\cdot 5\cdot 3y^{4+1}z^{6+2}$
$=75y^5z^8$

**29.** $V=LWH=(2y^4)(3y^2)(5y)=30y^7$

**30.** $\frac{57}{7} = 8\frac{1}{7}$

**31.** $3\frac{1}{5} = \frac{5\cdot 3+1}{5} = \frac{15+1}{5} = \frac{16}{5}$

**32.** $\frac{72\cancel{y}}{117\cancel{y}z} = \frac{\cancel{9}\cdot 8}{\cancel{9}\cdot 13z} = \frac{8}{13z}$

**33.** $\frac{27}{-45} = -\frac{\cancel{9}\cdot 3}{\cancel{9}\cdot 5} = -\frac{3}{5}$

**34.** $\frac{18x^7}{42x^{10}} = \frac{\cancel{6}\cdot 3\cancel{x^7}}{\cancel{6}\cdot 7\cancel{x^7}\cdot x^3} = \frac{3}{7x^3}$

**35.** $(3^3)^2 = 3^{3\cdot 2} = 3^6 = 729$

**36.** $\frac{21}{49} = \frac{\cancel{7}\cdot 3}{\cancel{7}\cdot 7} = \frac{3}{7}$

**37.** $\frac{156 \text{ ft}}{12 \text{ hr}} = 13$ ft per hr

**38.** (a) $\frac{42}{14} = 3$ secretaries per lawyer

(b) $\frac{12}{3} = 4$ paralegals per lawyer

(c) $\frac{12}{3} = \frac{n}{80} \Rightarrow 3n = 960$

$n = 320$ paralegals

**39.** $\frac{24}{x} = \frac{8}{5} \Rightarrow 8x = 24(5) = 120 \Rightarrow x = 15$

check: $\frac{24}{15} \stackrel{?}{=} \frac{8}{5},\ \frac{8}{5} = \frac{8}{5}$

**40.** $\frac{1}{90} = \frac{5}{n} \Rightarrow n = 5(90) = 450$ mi

**41.** $\frac{\cancel{8}\cancel{y}}{\cancel{15}}\cdot\frac{\overset{2}{\cancel{30}}}{\underset{3}{\cancel{24}}\, y^{\cancel{2}}} = \frac{2}{3y}$

**42.** $\frac{9}{14} \div \frac{-72}{21} = -\frac{\cancel{9}}{\cancel{7}\cdot 2}\cdot\frac{\cancel{7}\cdot 3}{\cancel{9}\cdot 8} = -\frac{3}{16}$

**43.** $\frac{1}{3} \div 4 = \frac{1}{3}\cdot\frac{1}{4} = \frac{1}{12}$ lb in each container

**44.** $6x = 2\cdot 3\cdot x,\ 18x = 2\cdot 3^2 x,\ 27 = 3^3$

$LCM = 2\cdot 3^3 x = 54x$

**45.** $\frac{8}{y} - \frac{4}{y} = \frac{8-4}{y} = \frac{4}{y}$

**46.** $\frac{-5}{16}\cdot\frac{3}{3} + \frac{9}{24}\cdot\frac{2}{2} = \frac{-15+18}{48} = \frac{3}{48} = \frac{1}{16}$

**47.** $\frac{5x}{18}\cdot\frac{5}{5} - \frac{7x}{45}\cdot\frac{2}{2} = \frac{25x-14x}{90} = \frac{11x}{90}$

**48.** $12\frac{1}{7}\cdot\frac{3}{3} - 7\frac{13}{21} = 11\frac{24}{21} - 7\frac{13}{21} = 4\frac{11}{21}$

**49.** $3\frac{3}{7} \div 6\frac{2}{3} = \frac{24}{7}\cdot\frac{3}{20} = \frac{6\cdot\cancel{4}\cdot 3}{7\cdot\cancel{4}\cdot 5} = \frac{18}{35}$

**50.** $\left(\frac{3}{4}\right)^2 + \frac{1}{9} \div \frac{1}{3} = \frac{9}{16} + \frac{1}{9}\cdot\frac{3}{1}$

$= \frac{9}{16}\cdot\frac{3}{3} + \frac{1}{3}\cdot\frac{16}{16}$

$= \frac{27+16}{48} = \frac{43}{48}$

**51.** $\frac{3}{4}x = 27 \Rightarrow x = \frac{4\cdot 27}{3} = 4\cdot 9 = 36$

**52.** $1\frac{1}{3}-\frac{6}{9}=\frac{4}{3}\cdot\frac{3}{3}-\frac{6}{9}=\frac{12-6}{9}=\frac{6}{9}=\frac{2}{3}$ in.

**53.** $(-4y^2-5)-(2y^2+3y+2)+(-y^2-3)$
$=-4y^2-5-2y^2-3y-2-y^2-3$
$=-7y^2-3y-10$

**54.** $(-3x)(x-7)=-3x^2+21x$

**55.** $(z^5)(z^6-3z-5)=z^{11}-3z^6-5z^5$

**56.** $(y+7)(4y^2-4y+6)$
$=4y^3-4y^2+6y+28y^2-28y+42$
$=4y^3+24y^2-22y+42$

**57.** $\angle b=x,\ \angle a=x+40°,\ \angle c=3x$

**58.** $(y+5)(y-3)=y^2-3y+5y-15$
$=y^2+2y-15$

**59.** $(x-4)(4x+3)=4x^2+3x-16x-12$
$=4x^2-13x-12$

**60.** $4z+12x=4(z+3x)$

**61.** $9y^5z^7+3y^6z^3=3y^5z^3(3z^4+y)$

**62.** $2x^2y^2-8xy+12x^2y=2xy(xy-4+6x)$

**63.** $-15-24=2x+15-3x$
$-39=-x+15$
$x=15+39$
$x=54$

check: $-15-24\overset{?}{=}2(54)+15-3(54)$
$-39=-39$

**64.** $-\frac{16}{4}=-3(6z)-5z\Rightarrow-4=-18z-5z$
$23z=4\Rightarrow z=\frac{4}{23}$
check: $-\frac{16}{4}\overset{?}{=}-3\left(6\cdot\frac{4}{23}\right)-5\cdot\frac{4}{23}$
$-4=-4$

**65.** $-z=\frac{1}{4}\Rightarrow z=-\frac{1}{4}$
check: $-\left(-\frac{1}{4}\right)\overset{?}{=}\frac{1}{4},\ \frac{1}{4}=\frac{1}{4}$

**66.** $5y-9-9y=23\Rightarrow-4y=32\Rightarrow y=-8$
check: $5(-8)-9-9(-8)\overset{?}{=}23,\ 23=23$

**67.** $7a-5+3a=-9+6a$
$10a-5=-9+6a\Rightarrow4a=-4\Rightarrow a=-1$
check: $7(-1)-5+3(-1)\overset{?}{=}-9+6(-1)$
$-15=-15$

**68.** $-2(y-5)=6(y+5)-12$
$-2y+10=6y+30-12=6y+18$
$-8y=8\Rightarrow y=-1$

**69.** $\frac{x}{2}+\frac{x}{3}=10\Rightarrow3x+2x=60$
$5x=60\Rightarrow x=12$

**70.** (a) $L=$ fall students
$L+95=$ spring students
$L-75=$ summer students
(b) $L+(L+95)+(L-75)=386$
(c) $3L+20=386\Rightarrow3L=366$
$L=122$ fall students
$L+95=217$ spring students
$L-75=47$ summer students

**70.** (d) $122+(122+95)+(122-75)\stackrel{?}{=}386$
$386=386$

**71.** $751.7596=751.76$ nearest hundredth

**72.** $0.588+75+8.59=84.178$

**73.** $0.042\times 0.04=0.00168$

**74.** $40(9.50)+12(13.65)=\$543.80$

**75.** $-4.75\div 3.2=-1.484375$
$=-1.48$ nearest hundredth

**76.**

| Fraction | Decimal | Percent |
|---|---|---|
| $\frac{1}{4}$ | 0.25 | 25% |
| $\frac{14}{25}$ | 0.56 | 56% |
| $3\frac{1}{2}$ | 3.5 | 350% |

**77.** $79=0.25n\Rightarrow n=\frac{79}{0.25}=316$

**78.** $0.30(90)=27$

**79.** $\frac{5.75}{28.50}\approx 0.20175$
$=20\%$ nearest hundredth

**80.** $100\%-60\%=40\%$

**81.** (a) $7(50)-6(50)=50$ students
(b) $4(50)+7(50)=550$ students

**82.** (a) $40\%+30\%=70\%$
(b) $0.10(2600)=\$260$

**83.**

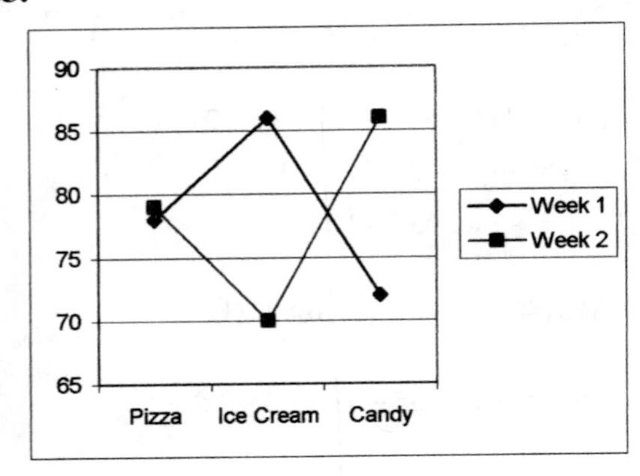


**84.** 59, 69, 72, 77, 78, 83, 87, 92, 95
median = 78

**85.** $\frac{89+87+92+97+85+85+95}{7}=90°\text{F}$

**86.** 2, 7, 7, 8, 8, $\underbrace{9, 9, 9}_{\text{mode}}$, 10

**87.** (a) $R(-3,5)$
(b) $T(0,3)$
(c) $S(-4,-4)$
(d) $U(6,2)$

**88.** $y=2x-4\Rightarrow x=\frac{y+4}{2}$

$x=\frac{-6+4}{2}=-1\Rightarrow(-1,-6)$

$x=\frac{-4+4}{2}=0\Rightarrow(0,-4)$

$x=\frac{-8+4}{2}=-2\Rightarrow(-2,-8)$

**89.** $y = 3x - 2$

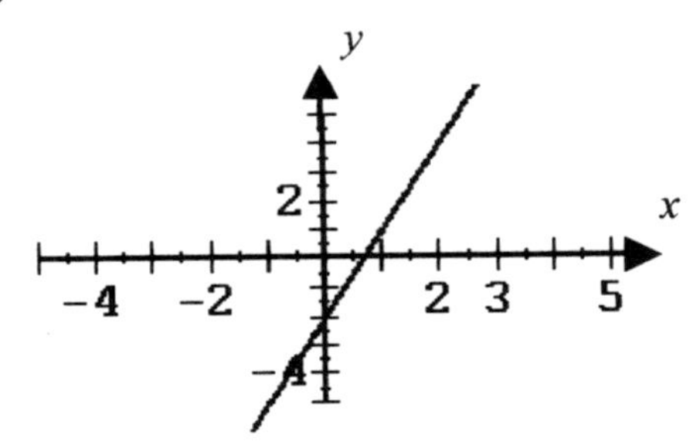


**90.** $y = -3$

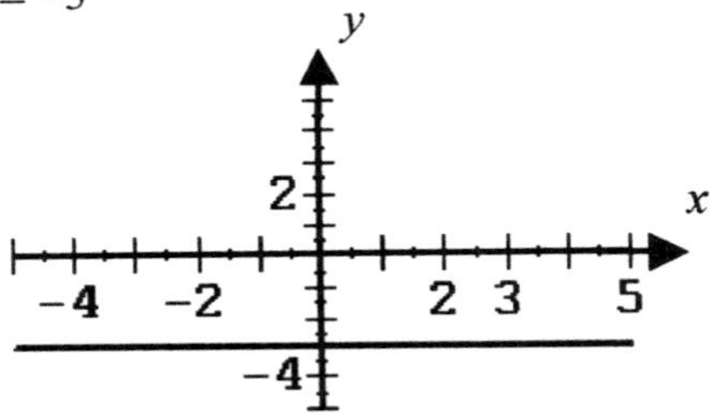


**91.** $14 \text{ gal} \cdot \frac{4 \text{ qt}}{\text{gal}} = 56 \text{ qt}$

**92.** $8 \text{ km} = 8000 \text{ m}$

**93.** $13 \text{ ft} \cdot \frac{0.305 \text{ m}}{\text{ft}} = 3.965 \text{ m}$

**94.** $\sqrt{81} = 9$

**95.** $\sqrt{121} + \sqrt{16} = 11 + 4 = 15$

**96.** $26^2 = 24^2 + x^2 \Rightarrow x^2 = 676 - 576 = 100$
$x = 10 \text{ in.}$

**97.** $C = \pi d \approx 3.14(13) = 40.82 \text{ ft}$

**98.** $A = \pi r^2 \approx 3.14\left(\frac{18}{2}\right)^2 = 254.34 \text{ m}^2$

**99.** $V = \frac{Bh}{3} = \frac{8(10)(9)}{3} = 240 \text{ ft}^3$

**100.** $\frac{7}{3} = \frac{49}{n} \Rightarrow 7n = 3(49) \Rightarrow n = 21 \text{ m}$